HOME WORKSHOP SILENCERS I

HOME WORKSHOP SILENCERS I

Published By Paladin Press,
Boulder, Colorado

Home Workshop Silncers I

ISBN 0-87364-193-0
Printed in the United States of America

Published by Paladin Press, a division of
Paladin Enterprises, Inc., P.O. Box 1307,
Boulder, Colorado 80306, USA.
(303) 443-7250

Direct inquiries and/or orders to the above address.

WARNING

It is against the law to manufacture a firearm silencer without an appropriate license from the federal government. There are also state and local laws prohibiting the possession of such devices in many areas. Severe penalties are prescribed for violations of these laws. Be warned!

CONTENTS

PUBLISHER'S FOREWORD

The history of firearms and related equipment is most notable for its constant technological advancement. From the day that man created his first crude muzzle loading firearm, other men sought to improve its design. To this day amateur inventors and machinists work at improving current firearm designs. In Philippine huts, in garage workshops, in sand-blown tents, men still improvise and create new firearm variants. Some make the weapons to fight. Others because of a desire to invent something new.

Hiram P. Maxim, called the father of the firearm silencer, was a man with an insatiable need to invent. Firearms were his specialty. Much has been written about Maxim already, and it is not my purpose to present his story here. Suffice it to say that this man of high moral character and unimpeachable credentials felt the desire to create a silencing device for a firearm. He succeeded admirably. Importantly enough, the American government did not restrict his efforts.

Today the situation has changed in the United States. Home workshop inventors are saddled with a plethora of local, state, and federal laws that regulate firearm manufacture and experimentation. The most severe laws of all pertain to silencer ownership.

When Hiram Maxim began selling his silencers around 1908, most buyers were solid citizens who wanted only to shoot their guns *quietly*. Their reasons were logical: Ideal for indoor target ranges, didn't scare the neighbors, good for pest control amidst farm animals, etcetera.

For over 20 years, until 1934, silencers were perfectly legal to own and use in America. Many long guns were threaded at the factory to accept standard Maxim silencers.

Then, in response to the notorious gang warfare of the twenties, the U.S. government passed the National Firearms Act of 1934. This law permanently controlled the private ownership of automatic weapons and silencers. Both items were commonly found in the gang arsenals of the day.

Currently firearm silencers are used in the U.S. primarily by law enforcement agencies and the military forces. It is interesting to note that private ownership of silencers is still quite legal in other civilized nations, such as Australia and France.

One of the main purposes of ***Home Workshop Silencers I*** is to help clear up the mystery surrounding firearm silencers. The designs presented here are precise and of highest quality. They are also relatively simple. After considering the designs, readers will see that any competent machinist can build a good silencer in his home workshop without the aid of this book. Mobsters in the twenties had armorers capable of producing good silencers, and they still have similarly competent armorers on their payrolls.

Criminals have always been able to buy silencers, as noted journalist and silencer expert J. David Truby has shown so clearly in his books ***Silencers, Snipers And Assassins*** and ***The Quiet Killers I & II.*** This has been the case since 1934, when silencers were made illegal, to the present day.

We have no reason to believe the situation will change. Papers throughout the world regularly carry stories announcing the confiscation of silenced and automatic weapons from criminal gangs. These range in quality from crude tape-and-tube silencers, to factory silenced Ingram M-10's. State and Federal laws appear to have little impact upon illegal trade of this sort.

Underground armorers of a different nature, on the other hand, produced silenced weapons for behind-the-lines operations against the Nazis in WW II. Most notable among these self-reliant machinists were the Dutch armorers. Their well-made silenced copies of the British Sten submachinegun took a heavy toll among the Nazis who invaded their country. Guerrilla and counterguerrilla forces around the globe continue to rely on the silenced submachinegun as one of their primary mission weapons.

Unfortunately, America's pop-culture-consuming public is particularly ignorant about firearm silencers. They connect the devices with B-grade spy films involving violence and assassination, or cheap novels of the same ilk. After considerable research, a Paladin editor has discovered what is believed to be the first piece of popular fiction involving a firearm silencer. This story is notable because it, and similar fiction which has followed, have forever tainted the silencer's public image.

Titled ***The Silent Bullet,*** the story was written in 1910 by Arthur Reeve, and features "Craig Kennedy, The Scientific Detective." Following are some pertinent excerpts from Reeve's rather florid tale, featuring the narrative of Kennedy:

> "Kerr Parker was surrounded by a group of people who were in his schemes with him. They are holding a council of war in the director's room. Suddenly Parker rises, staggers toward the window, falls, and is dead before a doctor can get to him. . . .
>
> "Now here comes the amazing part of the story. The doors to the offices on both sides were open at the same time. There were lots of people in each office. There was the usual click of typewriters, and the buzz of the ticker, and the hum of conversation. We have any number of witnesses to the whole affair, but as far as any of them knows no shot was fired, no smoke was seen, no noise was heard, nor was any weapon found. Yet here on my desk is a thirty-two calibre bullet. The coroner's physician probed it

out of Parker's neck this afternoon and turned it over to us."

The "scientific detective" wraps up his investigation:

> "I refer to this little device of a Hartford inventor. I place it over the muzzle of the thirty-two calibre revolver I have so far been using — so. Now, Mr. Jameson, if you will sit at that typewriter over there and write — anything so long as you keep the keys clicking. The inspector will start that imitation stock-ticker in the corner. Now we are ready. I cover the pistol with a cloth. I defy anyone in this room to tell me the exact moment when I discharged the pistol. I could have shot any of you, and an outsider not in on the secret would never have thought that I was the culprit. To a certain extent I have reproduced the conditions under which this shooting occurred . . ."

The Silent Bullet most probably is responsible for initiating the myth of the totally silenced firearm, as its title implies. The story abounds with glaring fallacies. A revolver cannot be satisfactorily silenced. And what about the muzzle flash? A silencer at best will only muffle a firearm's report, altering it, and quieting it somewhat. No inventor has ever created a completely effective silencer.

But they continue to try.

NOTE: Since Paladin Press has already published the books ***Home Workshop Guns For Defense And Resistance Volume I: The Submachinegun,*** and ***Vol. II: The Pistol,*** the silencer patterns presented here are designed to accommodate the two above weapons. However, the silencers can be adapted to any similar weapons.

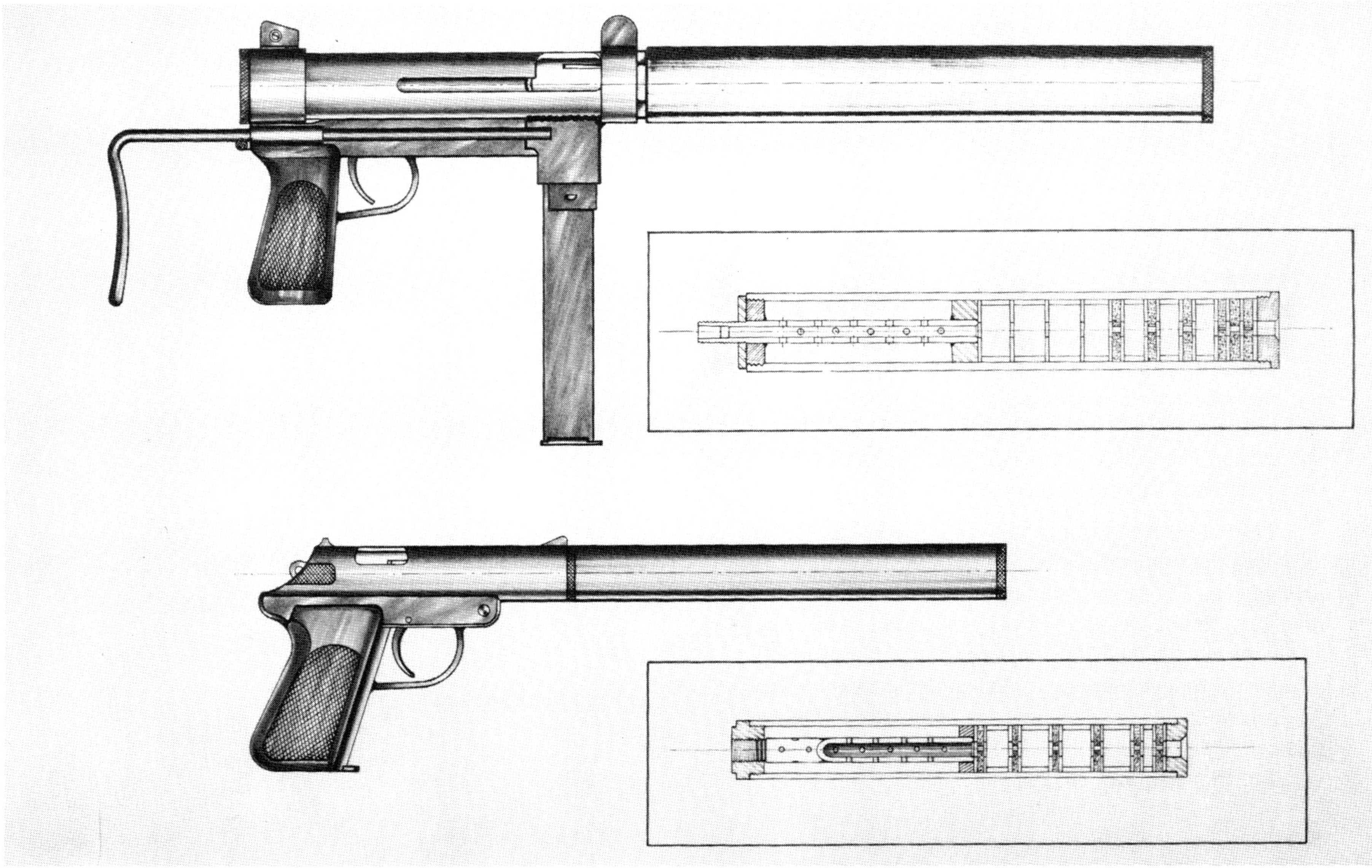

I. INTRODUCTION

As we know them today, firearm silencers are devices designed to muffle the discharge of a gun. Most conventional silencers simply look like long tubes 1 to 2 inches in diameter, and 8 to 15 inches long. An effective silencer usually looks awkward and heavy when attached in place to a firearm's muzzle. The thimble-size silencers one sees commonly in the cinema are the product of some prop director's imagination. To be functional, the laws of physics demand that a silencer must have the basic dimensions as outlined above. Inventors have designed and built smaller silencers than this. However, they are never as efficient as those of standard, albeit bulky, size.

Tracing back to the era from 1908 to 1934, we find that firearm silencers were available over-the-counter in many American hardware stores and gun shops. Most popular among the commercial silencers of the day was the Maxim Silencer. They were especially suited for use on .22 rifles. This set-up enabled the owner of a silenced .22 to shoot in his own backyard without annoying neighbors or farm animals.

After the lawless decade of the twenties, the American government realized what terror the silencer could produce when used by criminals. They quickly regulated silencer possession with the passing of the Firearms Control Act of 1934.

Today in the U.S. silencer ownerships is regulated by the Bureau of Alcohol, Tobacco, and Firearms (BATF). According to the BATF, a firearm silencer is a "Prohibited Device." To own a silencer, one must apply for a permit, be approved, then pay a $200 transfer tax. In detail the applicant must explain his specific reasons for wanting to own a silencer. The reasons must be undeniably legitimate.

Therefore, the reader must fully understand and be forewarned that before building or acquiring a silencer, he must first contact the proper authorities. The reader may only want to build a silencer as a machinist's challenge. He may have perfectly good reasons for owning one, and honorable intentions. Yet all this will matter not one bit if the BATF catches you with a silencer without having first obtained the proper permit. You will be fined, thrown into jail, and will never again be able to legally own ANY firearm. **Be warned!**

II. HISTORICAL & TACTICAL NOTES

Since its first appearance, the silencer's most important role has undoubtedly been with the military. Through both World Wars and to the present day, silenced firearms have earned a well-deserved place in the military arsenal.

In military applications, the silencer is generally used in conjunction with three specific types of firearms, each with its own special purpose. The first use is with a sniping rifle, as originated during WW I. Secondly, automatic pistols are silencer-equipped for covert purposes. Finally, silencers are fitted to submachineguns, generally for commando-type operations.

The first silenced submachineguns were developed during WW II, and included the Sten and the M3A1 "Greasegun". These weapons were created primarily to enable Allied commandos to quietly eliminate enemy sentries who guarded vital installations, such as weapons and gas depots. Resistance operations often utilized silenced smg's during this era as well.

A good silenced submachinegun can hit hard, fast, accurately, and quietly. They have added greatly to the success-potential factor where used behind the lines in resistance and guerrilla operations.

Arms designers and inventors around the world are busier today than ever before. At the vanguard of the firms manufacturing silenced weapons is Heckler & Koch of West Germany. Their extensive line of special purpose weaponry for military and police application has set a new pace in terms of quality, efficiency, and popularity.

Because of the rise of organized terrorism, many law enforcement agencies have formed special operational units to combat these political killers. In the U.S., these units are commonly known as SWAT teams. During their training, tactical employment of silenced counter-sniper rifles and automatic weapons is given strong emphasis.

Such training paid off for the West Germans in 1975. The government sent an elite commando group to the Mogadishu Airpot for a high-risk hostage rescue operation, which is now somewhat legendary. The operation was a complete success. No lives were lost, even though the hostages were held on board a jet. The West German commandos were armed with silenced Heckler & Koch MP5 submachineguns.

Today we have no reason to suspect that the popularity of silenced weapons will diminish in police and military circles.

III. MATERIALS & DIMENSIONS

Because I have no way of knowing exactly what type of materials and tools the reader intends to use, I must presume he has access to suitable tubing and tools. Some readers will have the finest grade tool steels, tubings, and even a full-blown machine shop to use. Others will have to scrounge piping and sheet steel from the local salvage yard, and have only the most basic of metal-working tools for his project. Realizing this, I have included plans for a very advanced submachinegun silencer which is intended to fill the needs of the skilled machinist. For those lacking the necessary skills and tools, I have created another version of the smg silencer, which is much less complicated to make. It is truly a simple, expedient design that works well in practice. As you might expect though, the simplified silencer is not as durable or attractive as the advanced one.

Also, dimensions presented here can be altered to suit your materials and tools. Yet the inside diameter of the smg silencer must not be less than 1-3/4 inches. Ideally, its inside diameter should measure 2 inches, with a length of at least 15 inches. This large interior area is necessary to contain expanding gases. You may think that a finished silencer of this size looks bulky. But remember that quietness is our main objective, so the silencer's interior area needs to be large.

In the case of the pistol silencer, the same rules apply. Its inside diameter can be from 1 to 1-1/4 inches, with a minimum length of 8 inches. I recommend the 1 inch diameter tubing here, since this size does not block the pistol's front sight.

If possible, use quality steel of known analysis for the tubings. Salvaged material such as shock absorber housings work well. However, other materials like brass or aluminum can also be substituted for the steel if nothing else is available for use as the silencer's outer housing. Even water pipes will do. Just be careful to lap down any interior seams the water pipe may have until its interior is uniformly smooth.

Remember that silencers are not as easily built as you may now think. They are nearly as complicated as any locked-breech firearm designed to handle assorted types of powerful ammunition. High pressures and velocities must be accommodated. Functioning must be reliable and safe. Needless to say, the silencer must be tested for all of these factors before it can be considered finished.

I also remind you that the muzzle velocity of the cartridge used in conjunction with a silencer should be subsonic (less than 1190 f.p.s.) for best results. Otherwise a supersonic "crack" will result from the speeding bullet.

In the case of the smg silencer, a ported replacement barrel is needed to reduce the velocity of the super-sonic 9 mm Parabellum cartridge.

Also you will note in the drawings that I refer to "Pairs" on baffle spacers and baffles. by "Pairs" I mean single units.

IV. THE SMG SILENCER

I recommend that the reader carefully study the drawings presented hereafter. It is much easier to glean the necessary dimensional information from them, than it is for me to try and write them down in book form. So if you are unclear on any given procedure or step, just refer to the appropriate machinist's drawing.

Tubing Construction Detail

As with building a firearm, the first part of the silencer to be made is the receiver housing. In a silencer design, this is more properly known as the outer tubing or housing.

The outer tubing contains all the parts that divert and contain expanding gases at the moment of cartridge ignition. By producing the outer tubing first, your job of making the finished silencer into a tight, well-made unit will be that much easier.

Your first step will be to obtain suitable tubing for the outer housing, as discussed earlier. It should be 15 inches long, with an inside diameter of 1.750/1.755 inches. Minimum thickness for the tubing is .200 inch for a thread allowance of 2.034 x 4-1/2 threads per inch (TPI). Thread both ends of the silencer tubing to these specifications, and refer to the drawings.

Some readers will not have access to large threading tools. For them I include a later chapter which presents a simplified silencer that needs no welding, brazing, or large threading tool. Some readers will therefore want to move ahead to that chapter at this point.

Smg Replacement Barrel

After the silencer's outer tubing is completed, a replacement barrel for the smg needs to be made. You may want to use the original barrel from your Home Workshop Smg. In choosing this approach, simply refer to the drawings for the data and dimensions you will need to modify the original barrel into a silenced barrel. Should you prefer to reserve your original barrel, build a new one according to the procedure presented in ***Home Workshop Guns I.***

The new silenced barrel will have the same specs as the original, with the exception of the following:

1. The threaded portion on the chamber end is reduced from 1-1/2 inches to 1-3/8 inches.
2. Gas ports are drilled into the barrel. Five 3/16 inch holes are first drilled through the barrel at 1 inch intervals. The second row of holes is staggered 1/2 inch longitudinally from the first row, 90 degrees away. Since holes are drilled through both barrel sides, there will be 20 holes in total. See drawing.
3. The new barrel is not rifled. It can be rifled if you wish, after the ports are drilled and the bore is lapped smooth. See ***HWG I*** for complete rifling instructions.

After the gas ports are drilled, look down the barrel. You will see burrs at every hole drilling point. These burrs must be removed completely, and the bore polished to a smooth finish. A stiff brass brush can be used initially to remove most of the burrs. Next use an ultra-fine emergy powder or similar compound for a final lapping of the bore.

Rear Silencer Plug

After the barrel is finished, the rear silencer plug is built. It can either be made from round stock of the specified diameter, or cut from 1/2 inch thick steel plate. If plate is chosen, expect a lot of work when curving it to a round shape unless you have a lathe. Because of this, I suggest round stock be used for the rear plug. The stock will be 2.034 inches in diameter before threading.

Do not cut the plug to its finished length of 1/2 inch **BEFORE** threading. There must be enough extra length in the round stock to allow it to be chucked in a lathe, if that is the threading procedure used. If a die is used for threading, the extra length will be needed to hold the round stock in a vise while it is threaded.

After the rear plug is properly threaded, cut it to its finished length.

As shown in the drawing, a hole .609 inch in diameter is centered within the plug. The barrel is press-fitted within this hole. That is why the hole's diameter is slightly smaller than the barrel diameter; for a snug fit. Some hand-fitting may be needed to make the press-fit here a tight and durable one.

Start the hole in the rear plug with a 3/16 inch drill. Gradually switch to larger drills until the final diameter of .609 inch is achieved. Be aware that this hole must be perfectly straight, since it will provide a base for the barrel and silencer tubing.

After the rear plug is press fitted into final place, it is brazed to the barrel. For brazing purposes, the front of the center hole in the rear plug is countersunk 60 degrees before press-fitting it in place. The plug is then brazed to the barrel in the countersunk area. The fillet that results from brazing will neatly fill the countersunk area, resulting in a truly professional job. Be sure to use high quality brazing rods, and don't "burn" the barrel when brazing. The burn will weaken the barrel, and it may soften or crack under the high pressures of actual use.

Welding may be substituted for brazing here. Just be careful, and be sure you know how to use the equipment properly.

Barrel End Support

The next part of the silencer to be built is the barrel end support. This part serves to support and align the barrel with the exit holes and baffles, as well as the other tubing.

Your primary concern from this point onward will be to precisely align the bore with all exit holes. Any alignment failure here will result in the bullets hitting the baffles or muzzle cap. The bullets will then break apart and lodge in the silencer. This could cause the silencer to blow-up in the hands of the person dumb enough to build and use such a shoddy unit. So take your time and be certain of the alignment before moving on to any steps that follow.

Use the same round stock for the barrel support that provided material for the rear plug. Its diameter should be 2.035 inches with a thickness of 3/4 inch. Note that the diameter of the barrel support is slightly larger than the inside diameter of the silencer tubing. This will make for a snug fit.

As with the rear plug, a hole .623/.625 in diameter is centered within the barrel support. It too will be press-fitted into place, and then brazed or welded to the barrel. Like the rear plug center hole, start the barrel support center hole with smaller drills, than gradually switch to larger ones until the final diameter is achieved.

To finish the end support, drill 4 holes into the face of the support at 90 degree points, as shown in the drawings. The four holes are 3/16 inch in diameter.

Actually these four holes serve as ports. They allow gases to circulate from the rear expansion chamber into the baffle chambers, located in front of the barrel support. These gases first propel the bullet, then are diverted by the barrel ports into the expansion chamber. From there they circulate forward through the four ports in the barrel support. They eventually dissipate in strength through this circulation and trapping system, and finally leak into the atmosphere through the baffle disks and front exit hole.

Before brazing the support to the barrel, be certain the end of the barrel is flush with the outside surface of the support. Finally, braze the countersunk area of the barrel support to the barrel surface according to the procedure given in the **Rear Silencer Plug** section.

Barrel Tightening Nut

You should now have a ported replacement barrel with rear plug and front support brazed into place. Screw the outer tube into the rear plug. Is the unit solid, with no play between the barrel support and the tube? Is every component perfectly aligned with the bore? If so, the next part to build is the barrel tightening nut.

Make this nut from round stock with a diameter of at least 2-1/8 inches. From the drawings we see that its thickness is 1/4 inch, and at its center is a 37/64 inch hole. Keep this hole straight and true, using graduated-sized drills for the work. When it is finished, thread it with a 5/8 x 18 TPI tap to match the barrel thread.

Next use a triangle file to serrate the outer surface of the tightening nut, to a depth of at least 1/16 inch. The serrations will allow for easier removal or tightening of the completed silencer. After filing, polish the serrations with emergy cloth to remove any edges or burrs. It may now be screwed into place at the end of the barrel.

Silencer End Cap

Here we have the last crucial part of the silencer assembly yet to be built. Like the other parts before it, care must be taken to insure precision of alignment. Perhaps most importantly, the hole in the center of the end cap must be perfectly aligned with the bore axis. I cannot stress this point strongly enough.

Start construction of the end cap by locating a piece of round stock preferably 2.200/2.250 inches in outer diameter (OD). Chuck the round stock in a lathe, and turn 3/8 inch of its length down to 2.034 OD. This area is threaded with a 2.034 x 4-1/2 TPI die, or turned with a lathe to these specs.

After threading the 3/8 inch length, add 1/4 inch of unthreaded, unturned stock, and cut for an overall length of 5/8 inch. Polish the front and back of the end cap until they are "square" and smooth.

Now here is a little trick to help insure alignment of the end cap's center hole with the bore axis. After the end cap is threaded and cut to length, screw it into the front of the silencer tubing. Also screw the barrel into the rear of the tubing. The unit should be solid with no play anywhere.

The alignment trick calls for a uniform length of drill rod with a diameter of .346/.350 inch. Its length should be about 20 inches, and it must fit snugly in the barrel bore. With the end cap in place, fit the drill rod into the rear of the barrel, pushing it forward until it touches the end cap. Put the silencer on a hard floor, with the end cap on the floor and the drill rod sticking straight up in the air. Gently tap the end of the drill rod with a hammer. Now remove the end cap from the silencer tubing. Its inside face should be faintly marked with the outline of the drill rod's diameter. Remember that this diameter is the same as the bore's.

The exit hole can now be drilled within the outline provided by the drill rod punch. Find the center of the outline's diameter, then drill the center hole first with a 3/16 inch bit. Keep changing drills until the final diameter of 3/8 inch is reached. Polish smooth.

Screw the end cap back into the silencer tubing. Insert the drill rod again into the barrel's rear until it reaches the end cap center hole. The rod should fit through the exit hole dead center, and must not touch any portion of the hole. If this is the case, you have done your job well.

Finish the end cap with a 45 degree countersink into the outside surface of the exit hole. This will aid the bullet in aligning itself, and help prevent lead shaving. The cap may then be knurled with a triangle file, polished with emergy cloth, and set aside.

Baffle Spacers

Baffle spacers are used to form channels or gas pockets in between the baffles. They trap the gases that propel the bullet out of the barrel. Because of this silencer's advanced design, most of these gases are trapped in the rear expansion chamber of the unit. Due to the four port holes drilled in the barrel support, gases first trapped in the expansion chamber circulate forward to the baffle channels, and vice-versa. Eventually the gases will escape through the front exit hole, via the baffles.

Construct the baffle spacers from steel, aluminum, brass, or even heat-proof plastic. This wide range of suitable materials is possible because the baffle spacers will not be subjected to extreme pressures since the outer tubing will support them. If you look hard enough, you should be able to locate a length of pipe that slip fits snugly inside the silencer tube. This pipe can be made of any of the suggested materials.

This length of pipe is cut precisely into three 7/8 inch length baffle spacers, three that are 1/2 inch long, and one spacer that is 3/4 inch in length. The tops and bottoms of these spacers must be flat, and "square" when compared to the spacer's height.

Baffle Supports

The baffle supports can be made of steel or bronze. They are merely disks with center holes that are the same diameter as the bore axis and end cap exit hole: 3/8 inch to 13/32 inch. The outer diameter of the baffle supports must fit snugly within the silencer tube's inner diameter. A thickness of 1/8 inch is ideal for the supports.

The baffle supports form channels to contain expanding gases. They also support the rubber baffle disks found at the front of the silencer (refer to diagram).

With any luck, you should be able to find some standard plumbing or machinist's washers at the hardware store to serve as the baffle supports.

In total, 13 baffle spacers are needed.

Baffle Disks

A variety of materials can be improvised to serve as the baffle disks. Hard rubber shoe soles will work, as will hockey pucks. Two or three pucks will provide all the material needed to make the baffles. If you desire the best possible material for the disks, locate some asbestos rubber at your local hardware store. Second best is neoprene rubber. Above all, the material selected for the baffle disks must be very durable.

Thickness of the disks can be 3/8 inch to 1/2 inch. The disks must fit tightly when assembled in the silencer tubing. From the drawings you will note that two metal baffle supports sandwich each baffle disk, holding the disk firmly in place when the bullet forces its way through the baffles and exit hole.

Six baffle disks are required.

Assembly And Testing

The silencer components can now be assembled according to directions on page 33. Your first test shots should be fired remotely behind some kind of barricade, in case the silencer malfunctions and blows up.

Heat Treatment And Bluing

After testing, any steel components of known analysis can be heat treated and then blued according to instructions found in the ***Home Workshop Guns*** books.

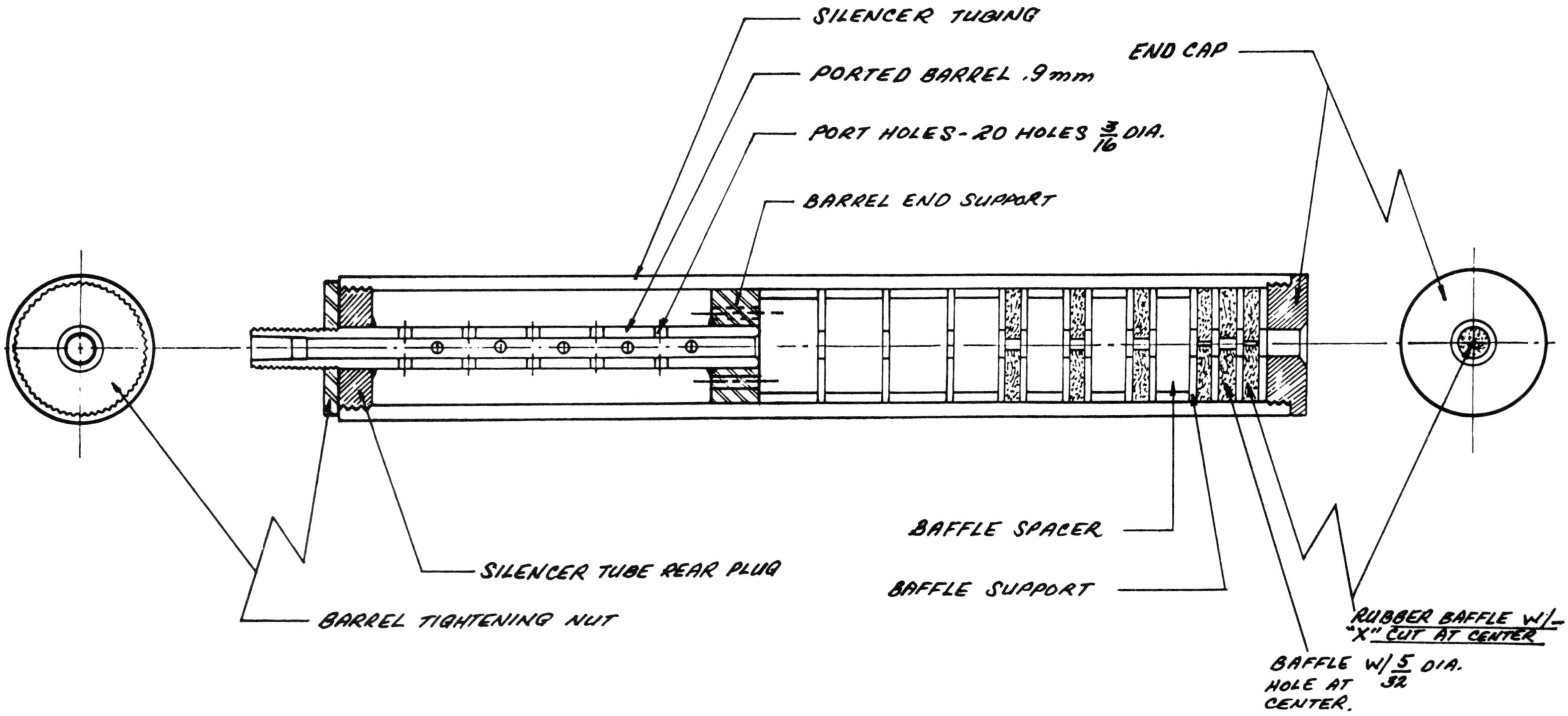

-SILENCER DETAIL - .9mm SUBMACHINEGUN-

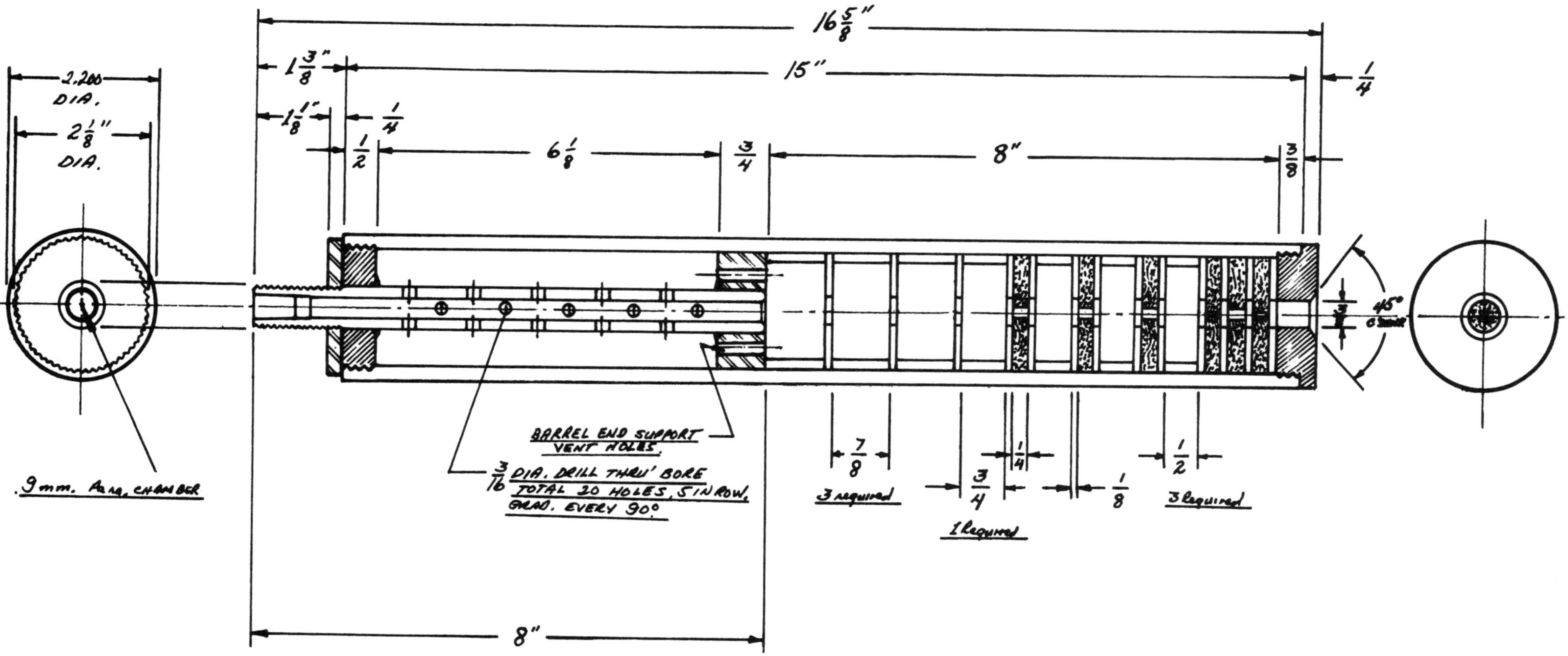
16 5/8"
15"
1 3/8"
1 1/8"
1/4
1/2
6 1/8
3/4
8"
3/8
1/4
2.200 DIA.
2 1/8" DIA.
.9mm. Para. CHAMBER
3/16 DIA. DRILL THRU' BORE
TOTAL 20 HOLES, 5 IN ROW,
GRAD. EVERY 90°
BARREL END SUPPORT
VENT HOLES.
7/8
3 required
3/4
1/4
1 Required
1/8
1/2
3 Required
45°
8"

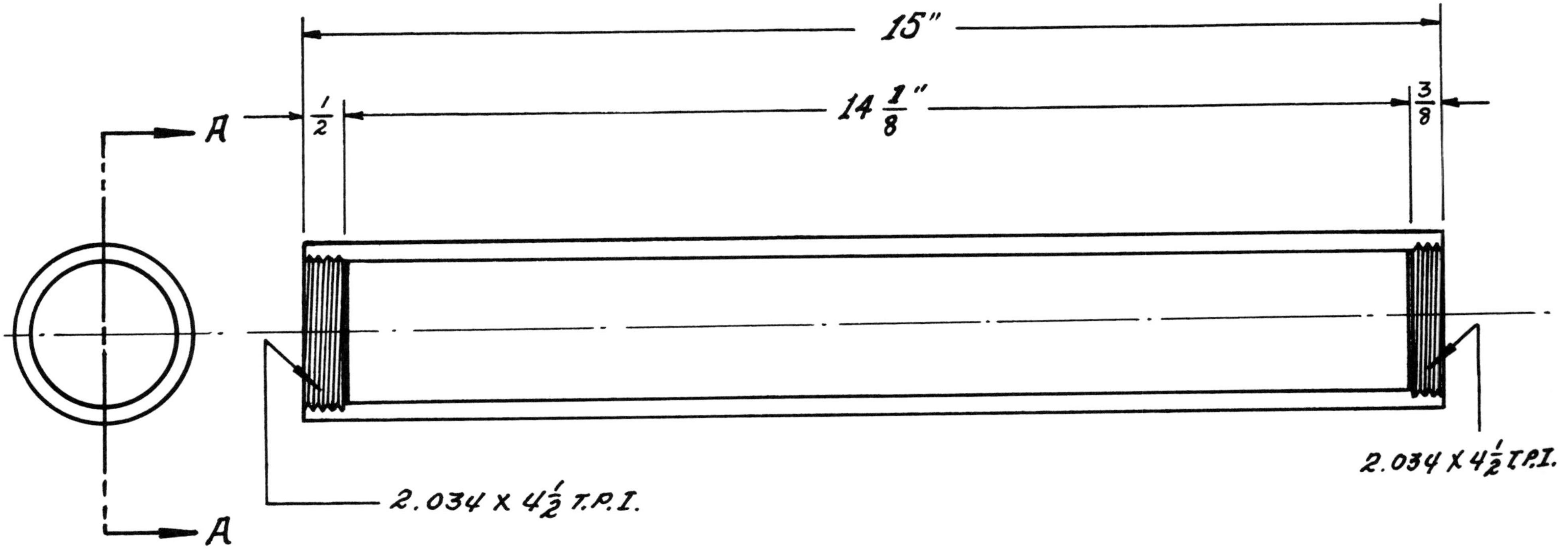

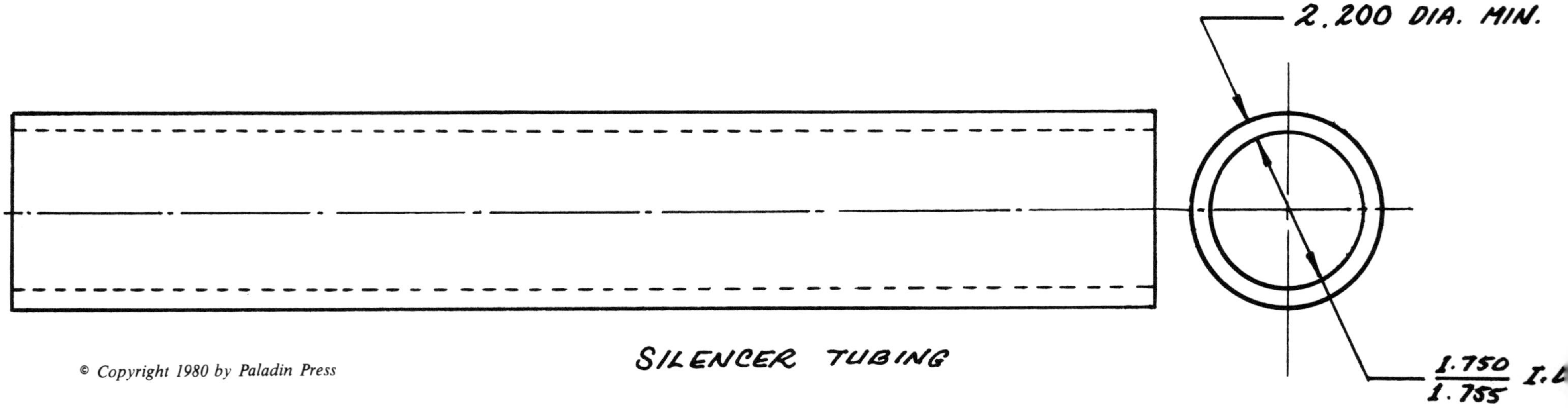

SILENCER TUBING

1

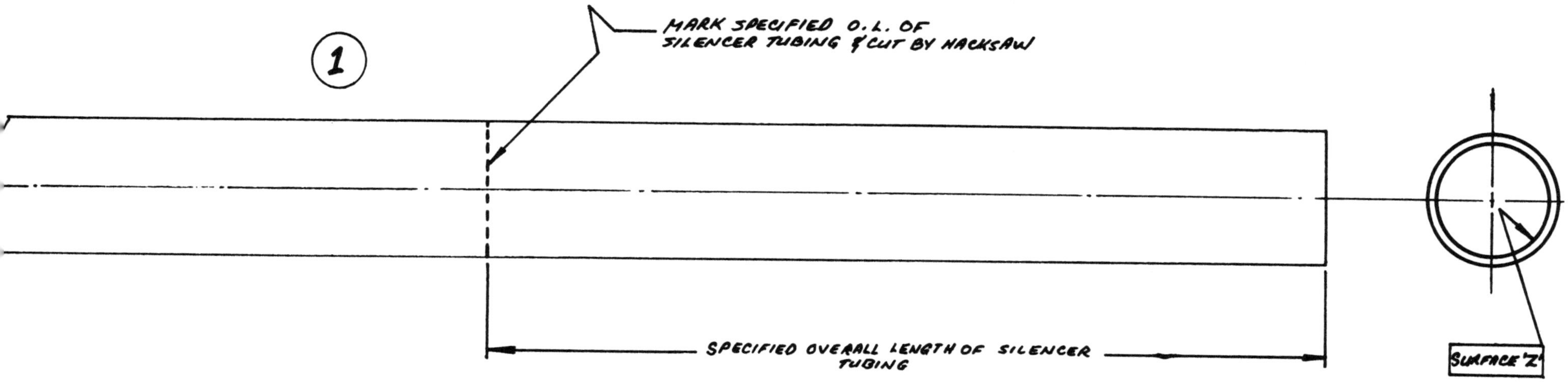

2

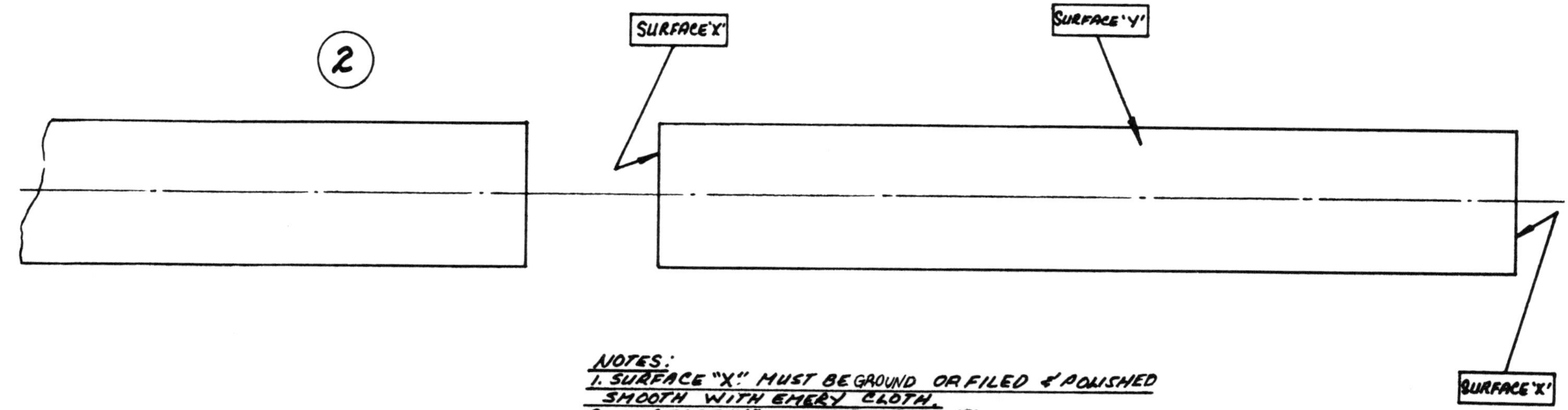

NOTES:

1. SURFACE "X" MUST BE GROUND OR FILED & POLISHED SMOOTH WITH EMERY CLOTH.
2. SURFACE "Y" MUST BE POLISHED WITH EMERY CLOTH & BUFFED.
3. SURFACE "Z" MUST BE POLISHED WITH EMERY CLOTH. WRAP A ROUND WOOD DOWEL WITH EMERY CLOTH & RUN BACK & FORTH INSIDE THE TUBING.

③ THREADING SILENCER TUBING ON VISE BY USING THREADING TOOL.

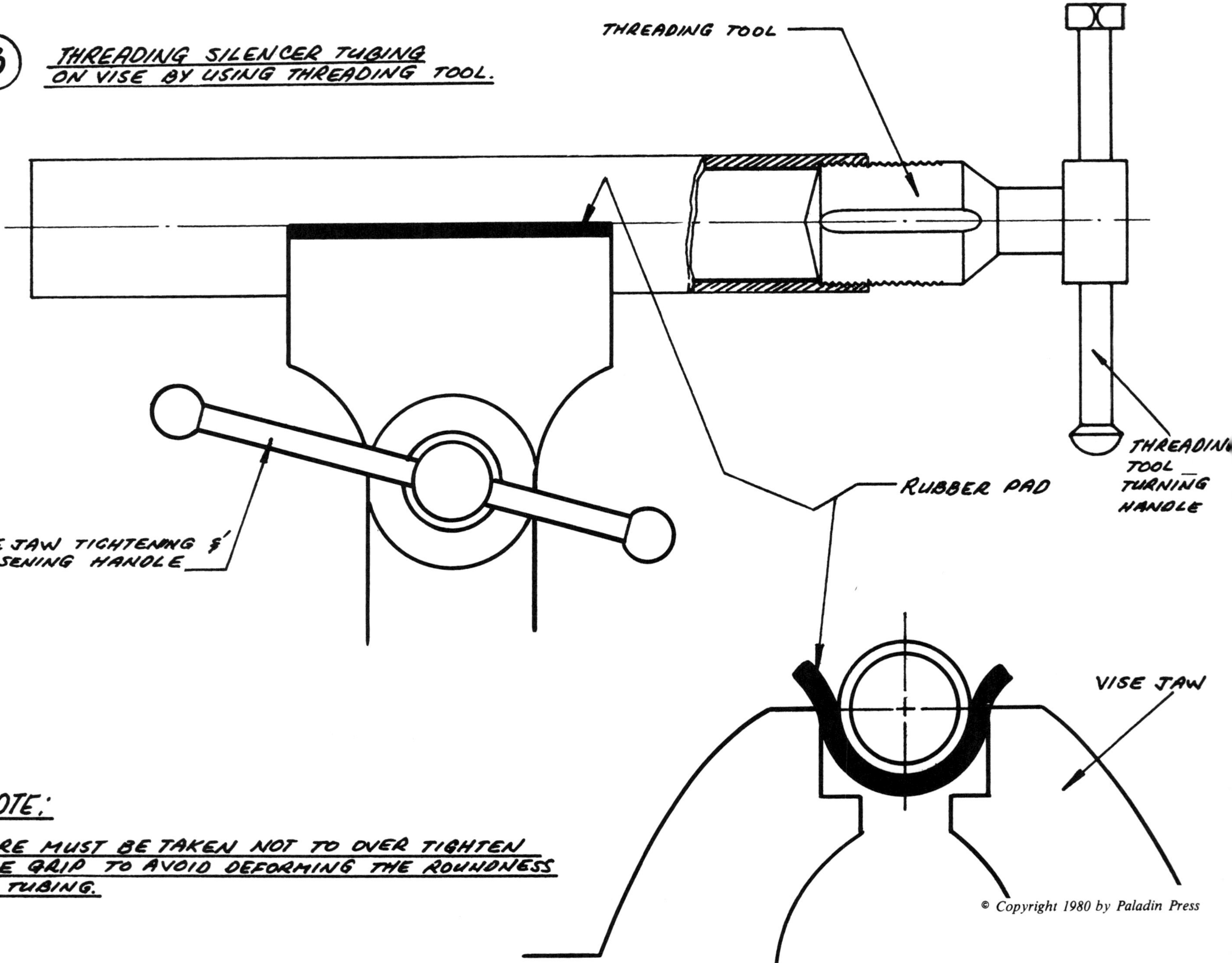

NOTE:

CARE MUST BE TAKEN NOT TO OVER TIGHTEN VISE GRIP TO AVOID DEFORMING THE ROUNDNESS OF TUBING.

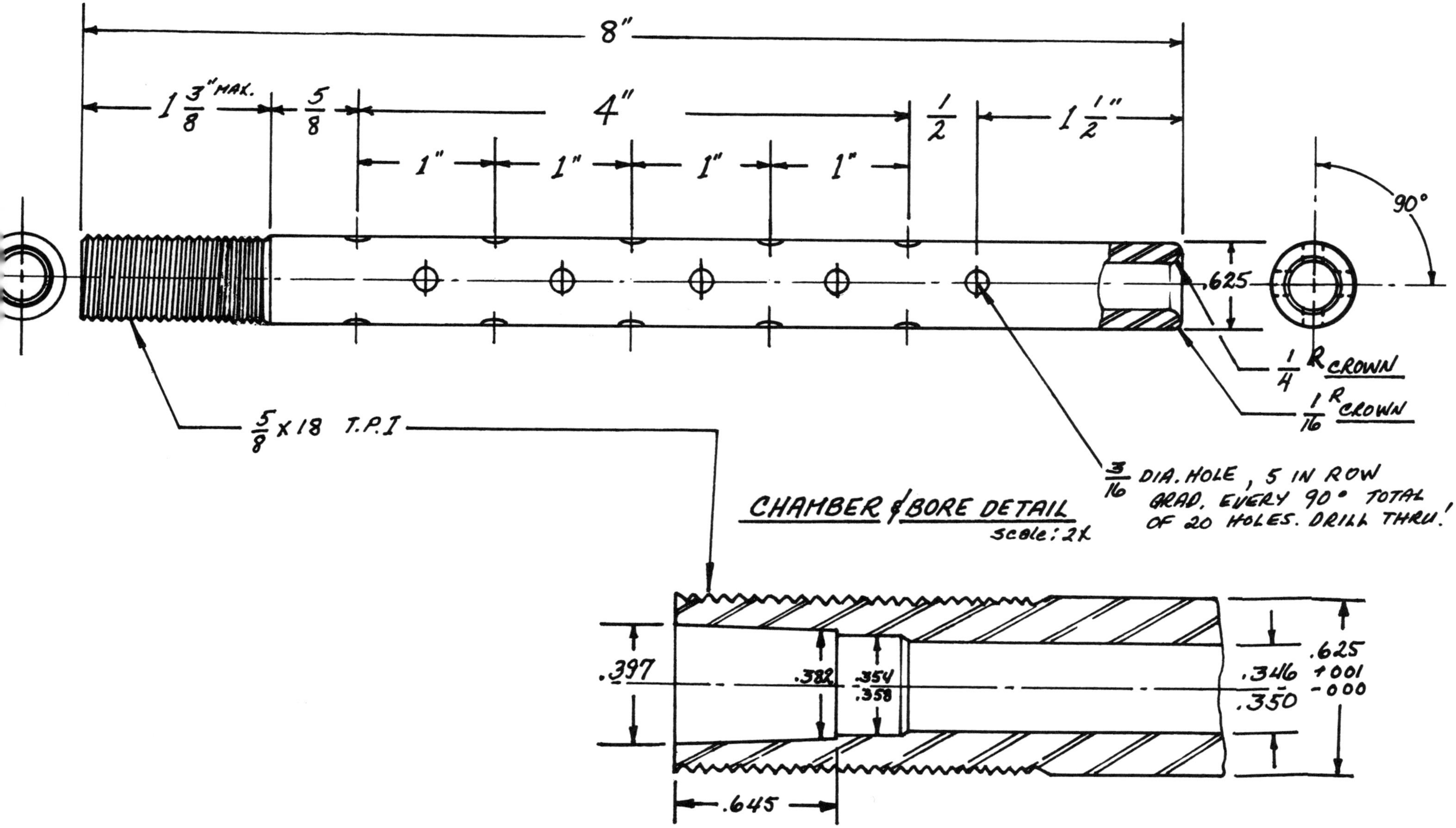

DETAIL: SILENCER BARREL. 9mm

SCALE
FULL

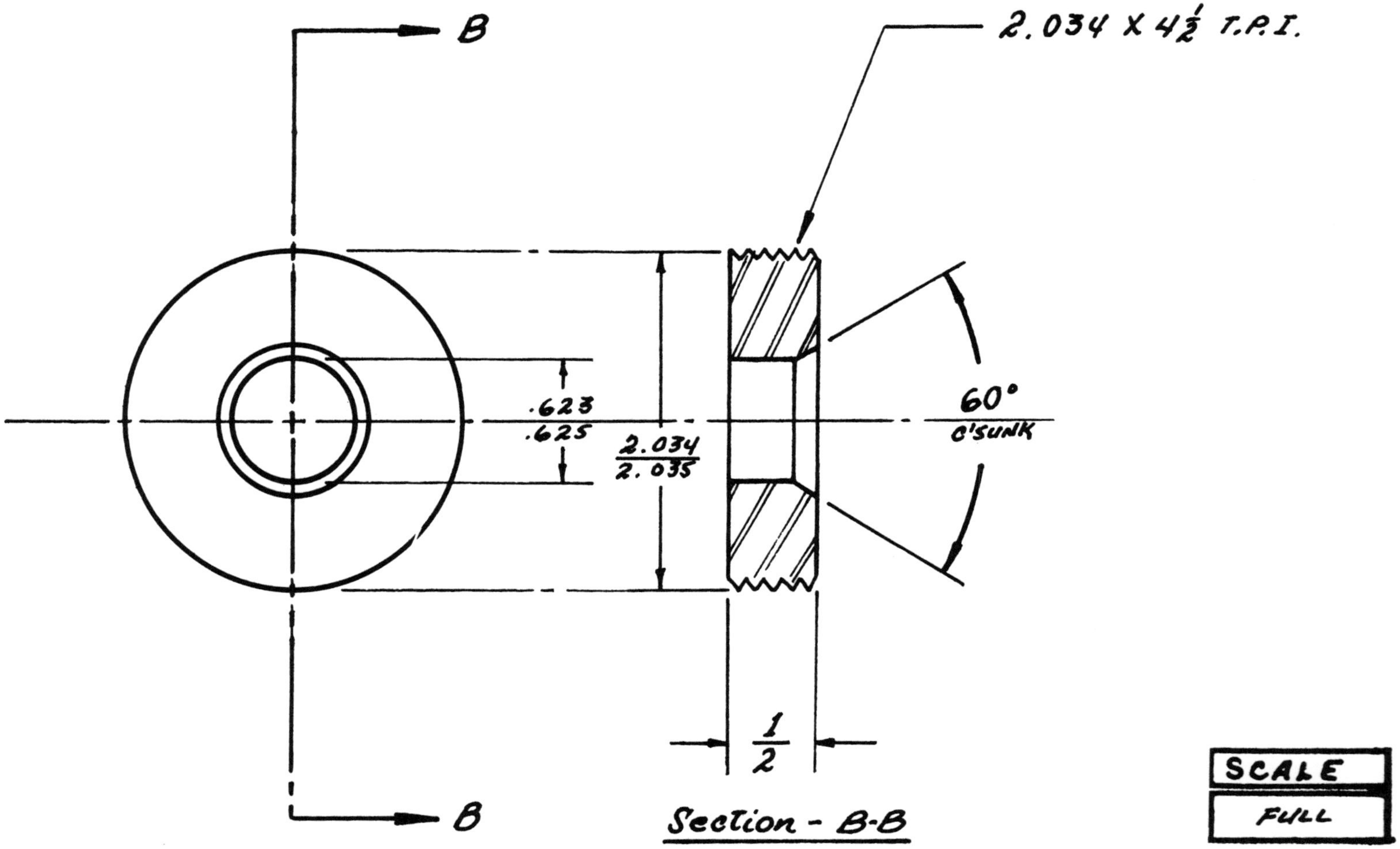

-REAR SILENCER PLUG-

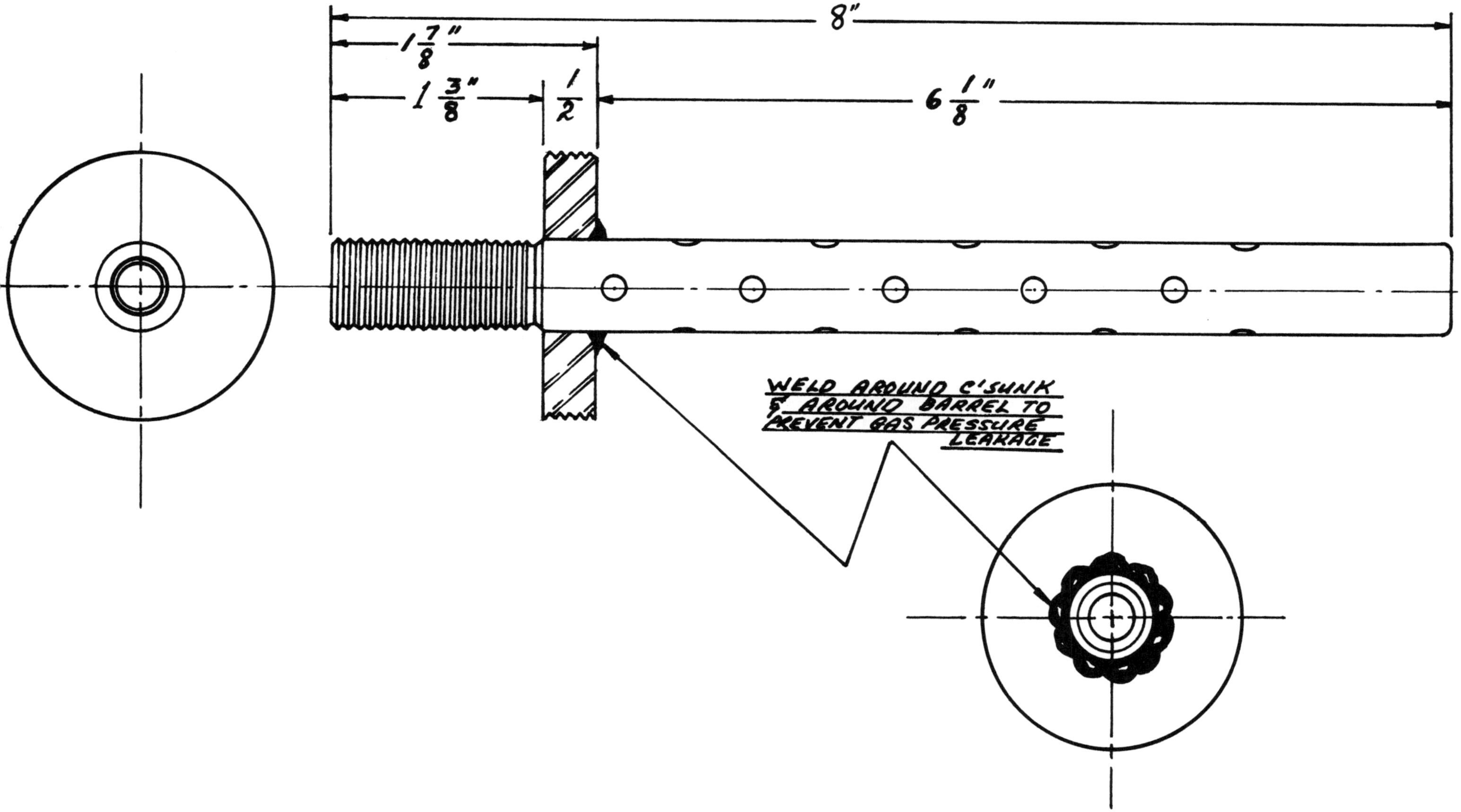

BARREL / REAR SILENCER CAP ASSY.

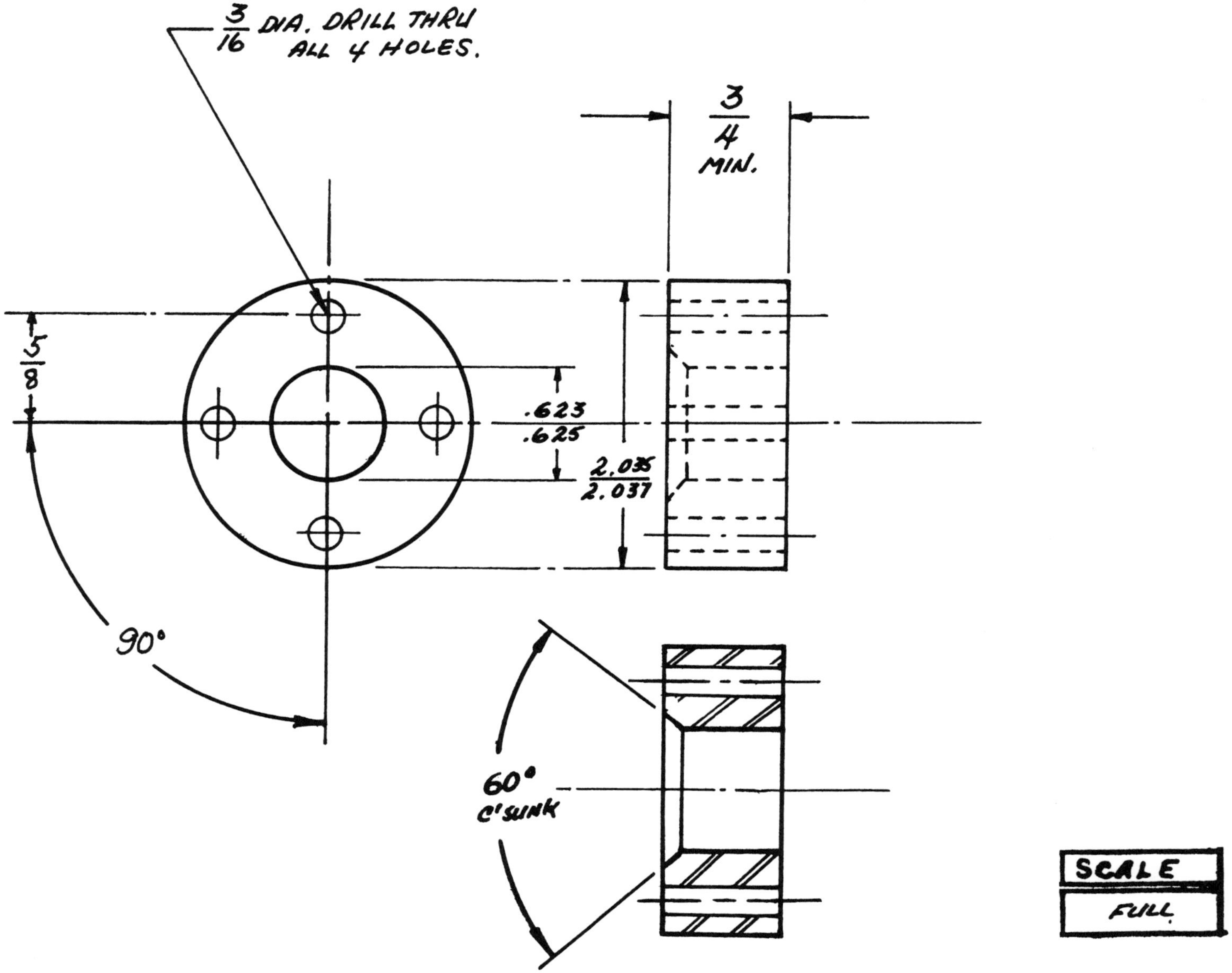

BARREL END SUPPORT

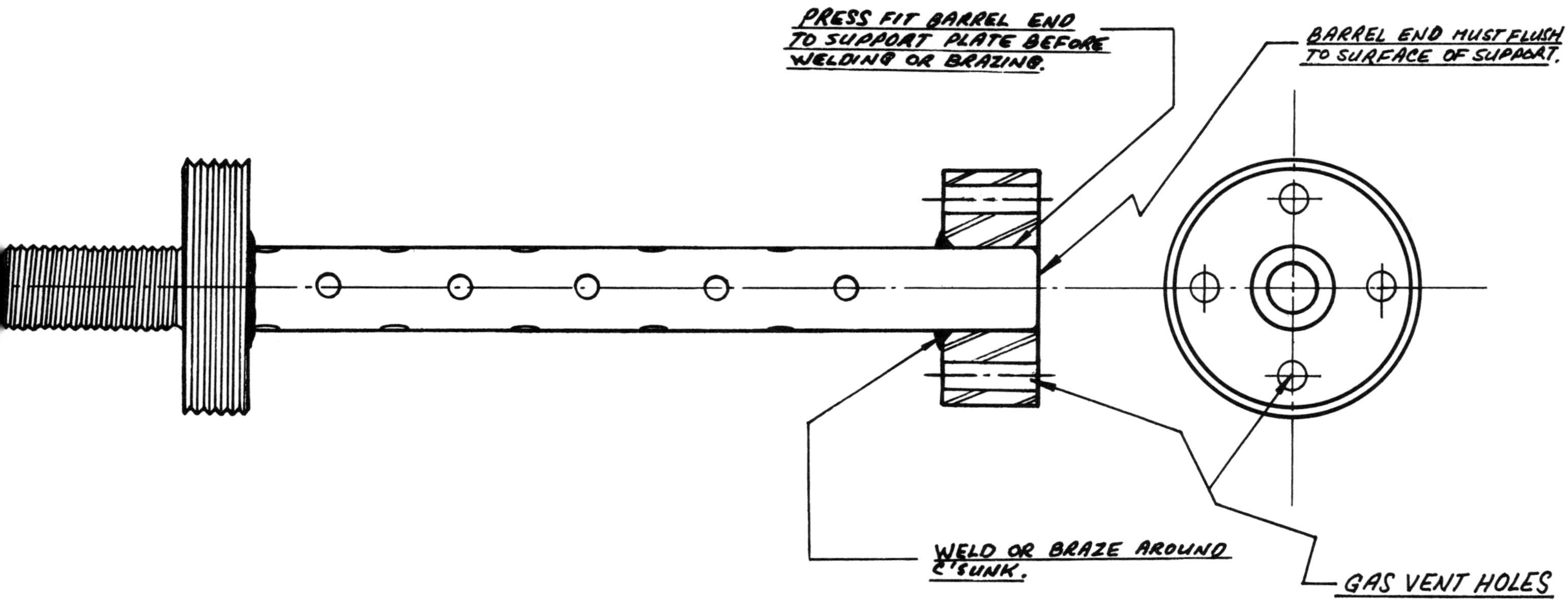

ASSEMBLY - BARREL & END SUPPORT PLATE

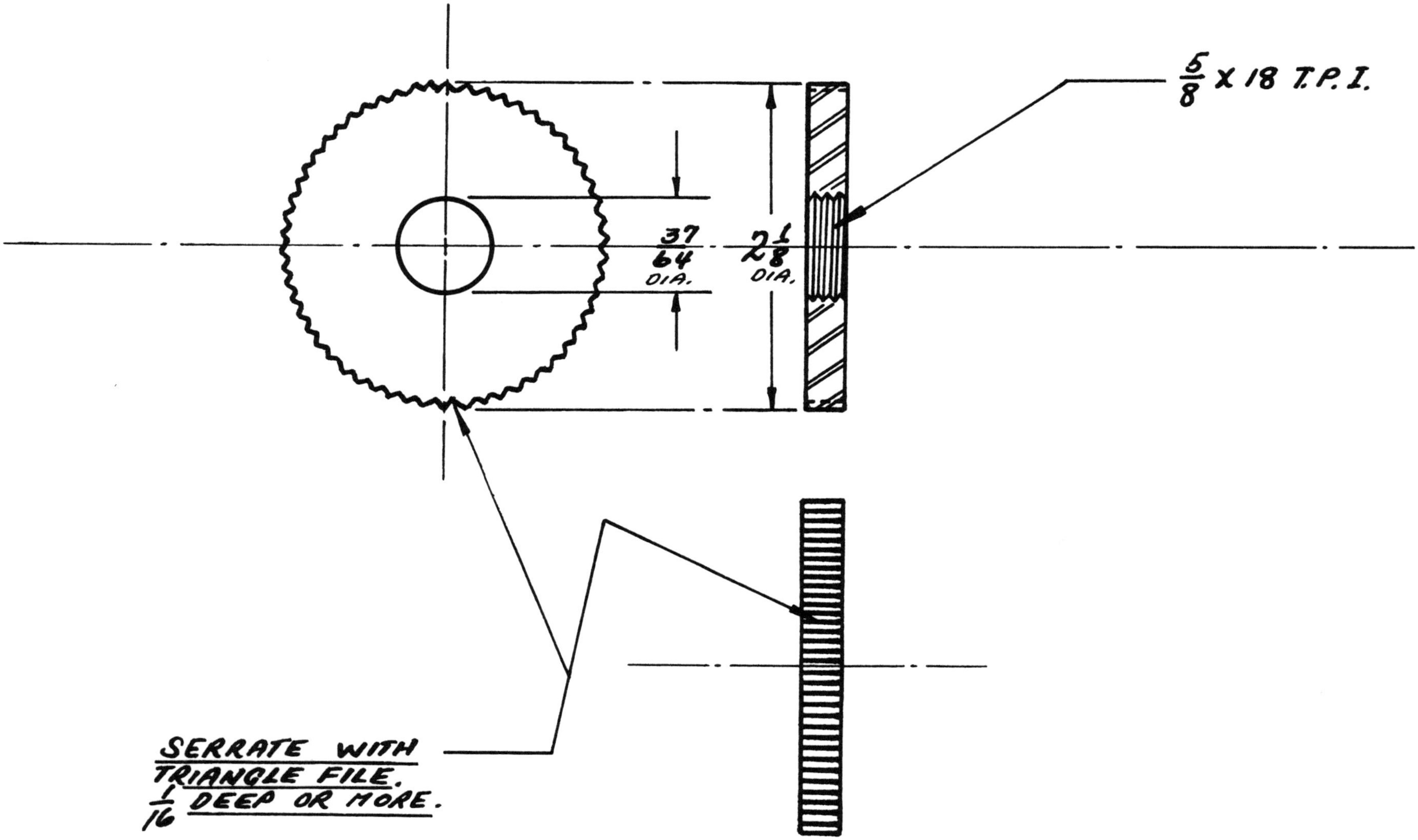

BARREL TIGHTENING NUT

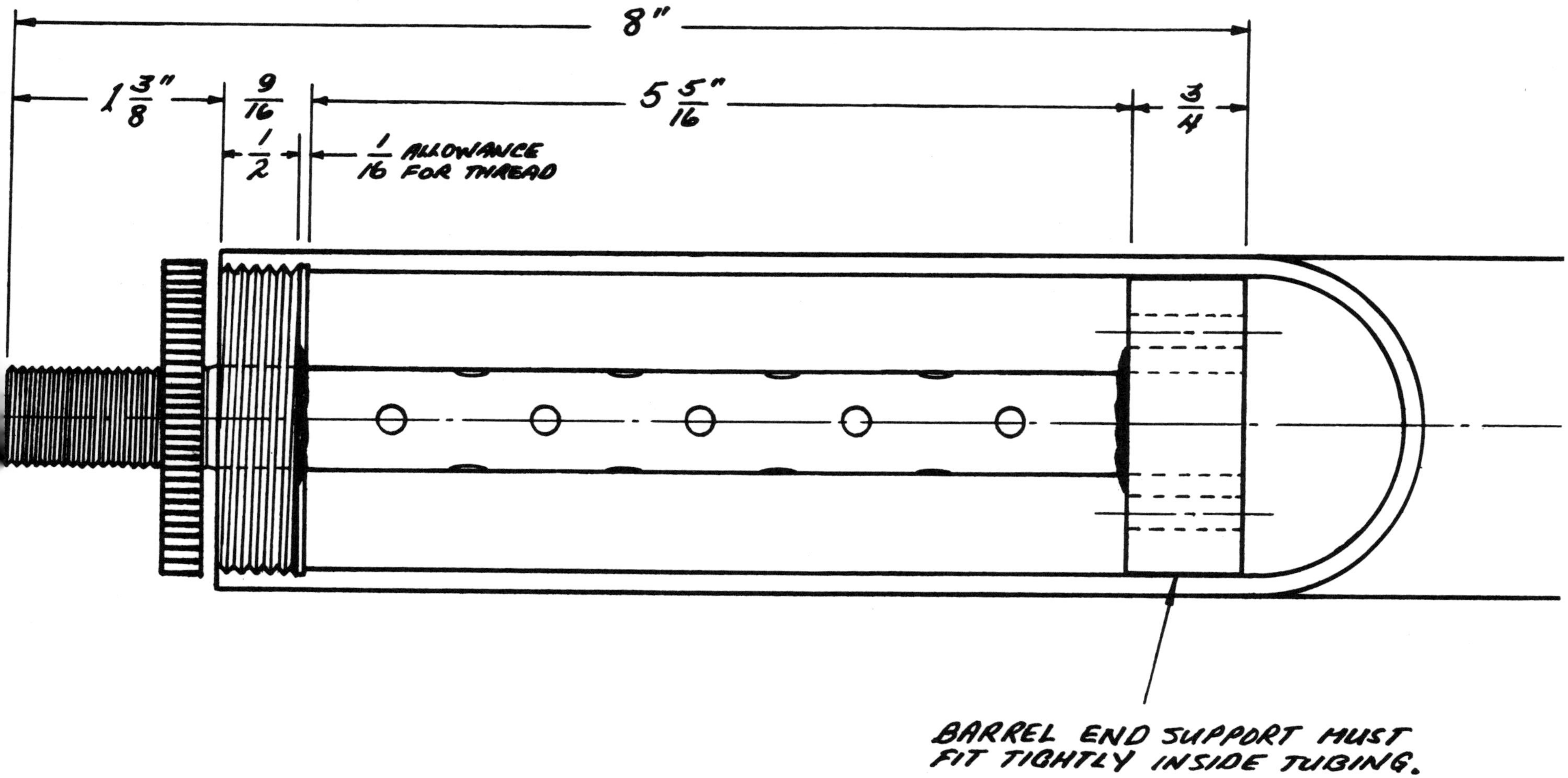

—BARREL-SILENCER TUBING ASSEMBLY—

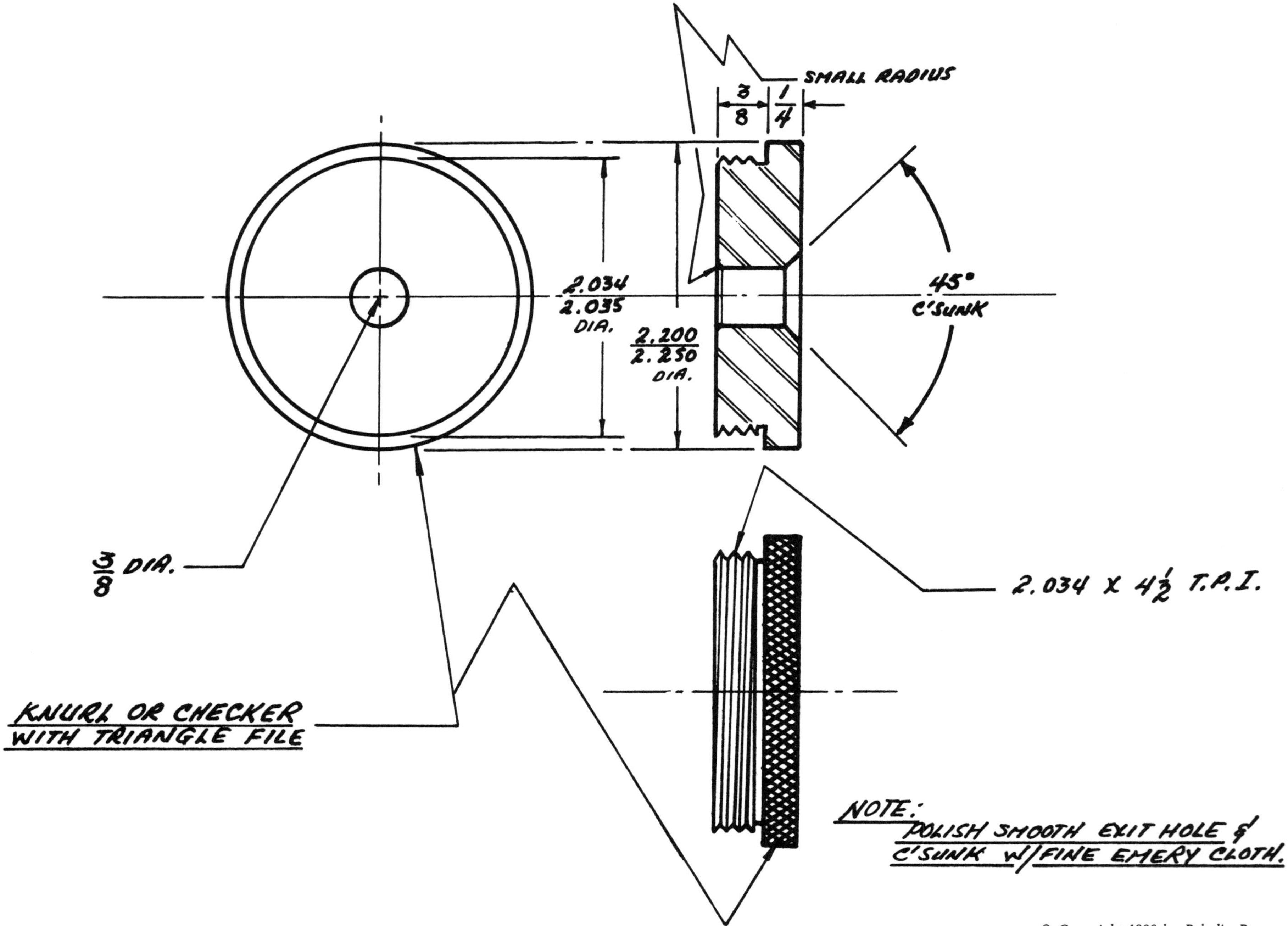

SILENCER END CAP

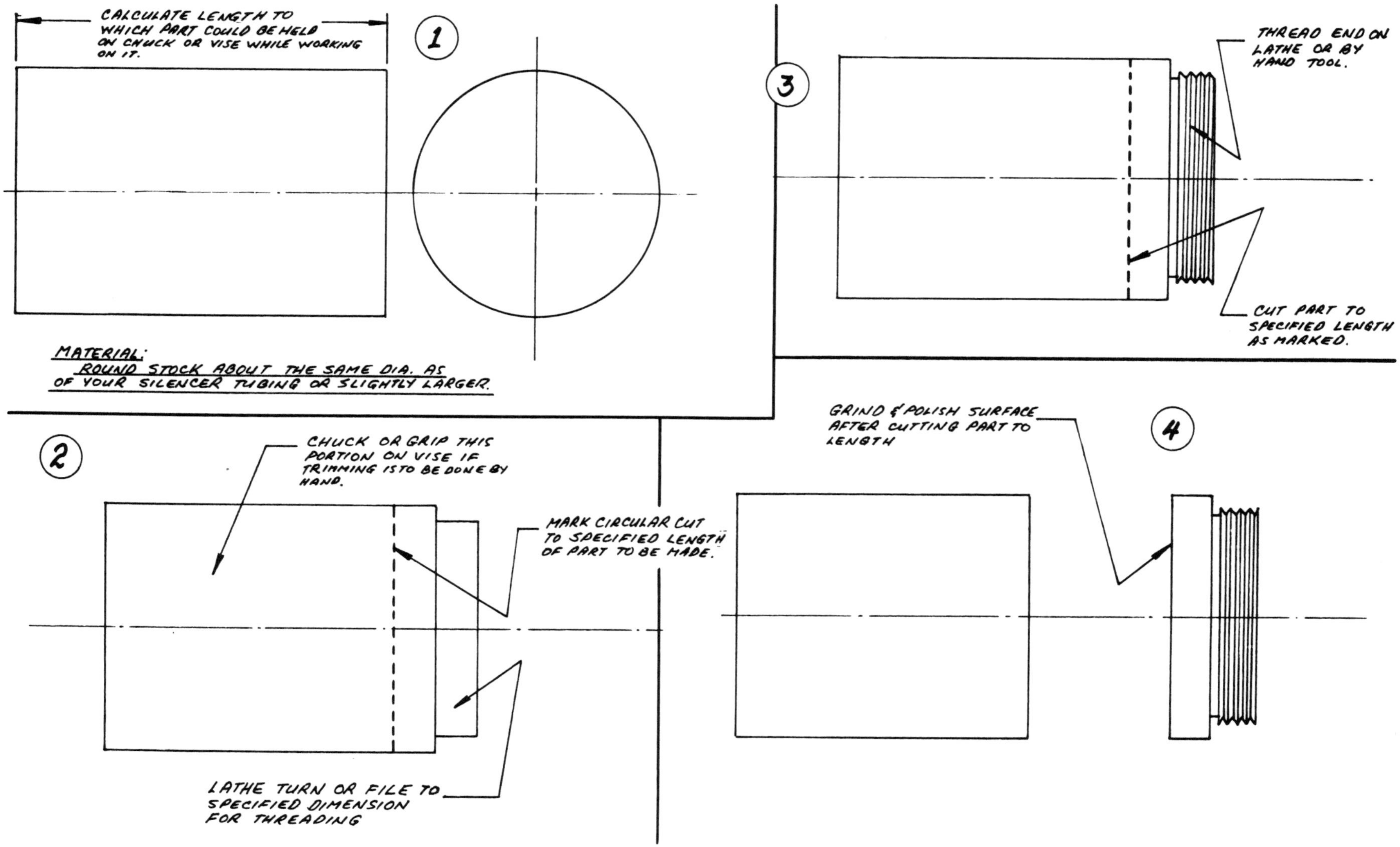
1
CALCULATE LENGTH TO WHICH PART COULD BE HELD ON CHUCK OR VISE WHILE WORKING ON IT.
MATERIAL:
ROUND STOCK ABOUT THE SAME DIA. AS OF YOUR SILENCER TUBING OR SLIGHTLY LARGER.
2
CHUCK OR GRIP THIS PORTION ON VISE IF TRIMMING IS TO BE DONE BY HAND.
MARK CIRCULAR CUT TO SPECIFIED LENGTH OF PART TO BE MADE.
LATHE TURN OR FILE TO SPECIFIED DIMENSION FOR THREADING
3
THREAD END ON LATHE OR BY HAND TOOL.
CUT PART TO SPECIFIED LENGTH AS MARKED.
4
GRIND & POLISH SURFACE AFTER CUTTING PART TO LENGTH

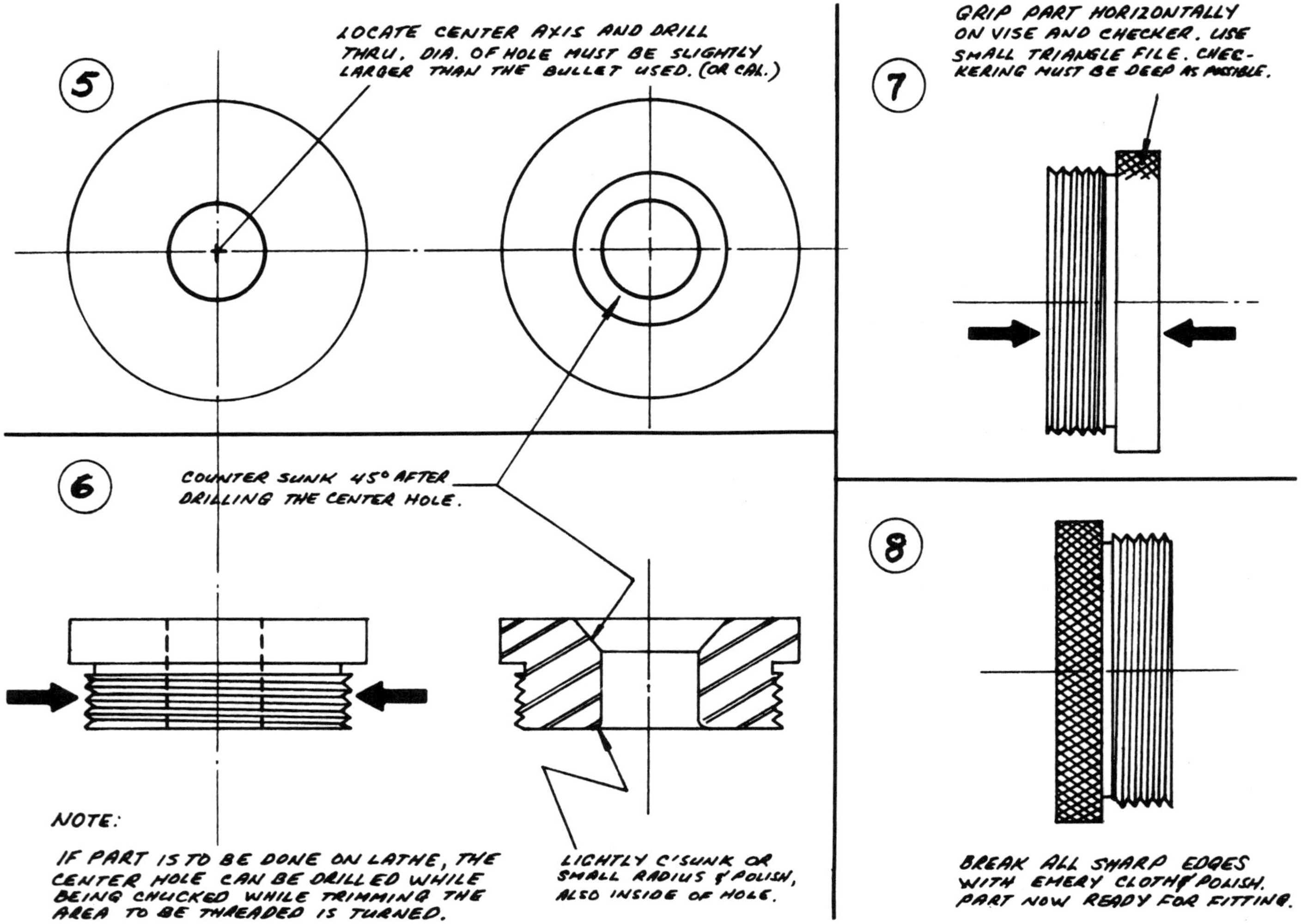
5
LOCATE CENTER AXIS AND DRILL
THRU. DIA. OF HOLE MUST BE SLIGHTLY
LARGER THAN THE BULLET USED. (OR CAL.)
7
GRIP PART HORIZONTALLY
ON VISE AND CHECKER. USE
SMALL TRIANGLE FILE. CHECKERING MUST BE DEEP AS POSSIBLE.
6
COUNTER SUNK 45° AFTER
DRILLING THE CENTER HOLE.
NOTE:
IF PART IS TO BE DONE ON LATHE, THE
CENTER HOLE CAN BE DRILLED WHILE
BEING CHUCKED WHILE TRIMMING THE
AREA TO BE THREADED IS TURNED.
LIGHTLY C'SUNK OR
SMALL RADIUS & POLISH,
ALSO INSIDE OF HOLE.
8
BREAK ALL SHARP EDGES
WITH EMERY CLOTH & POLISH.
PART NOW READY FOR FITTING.

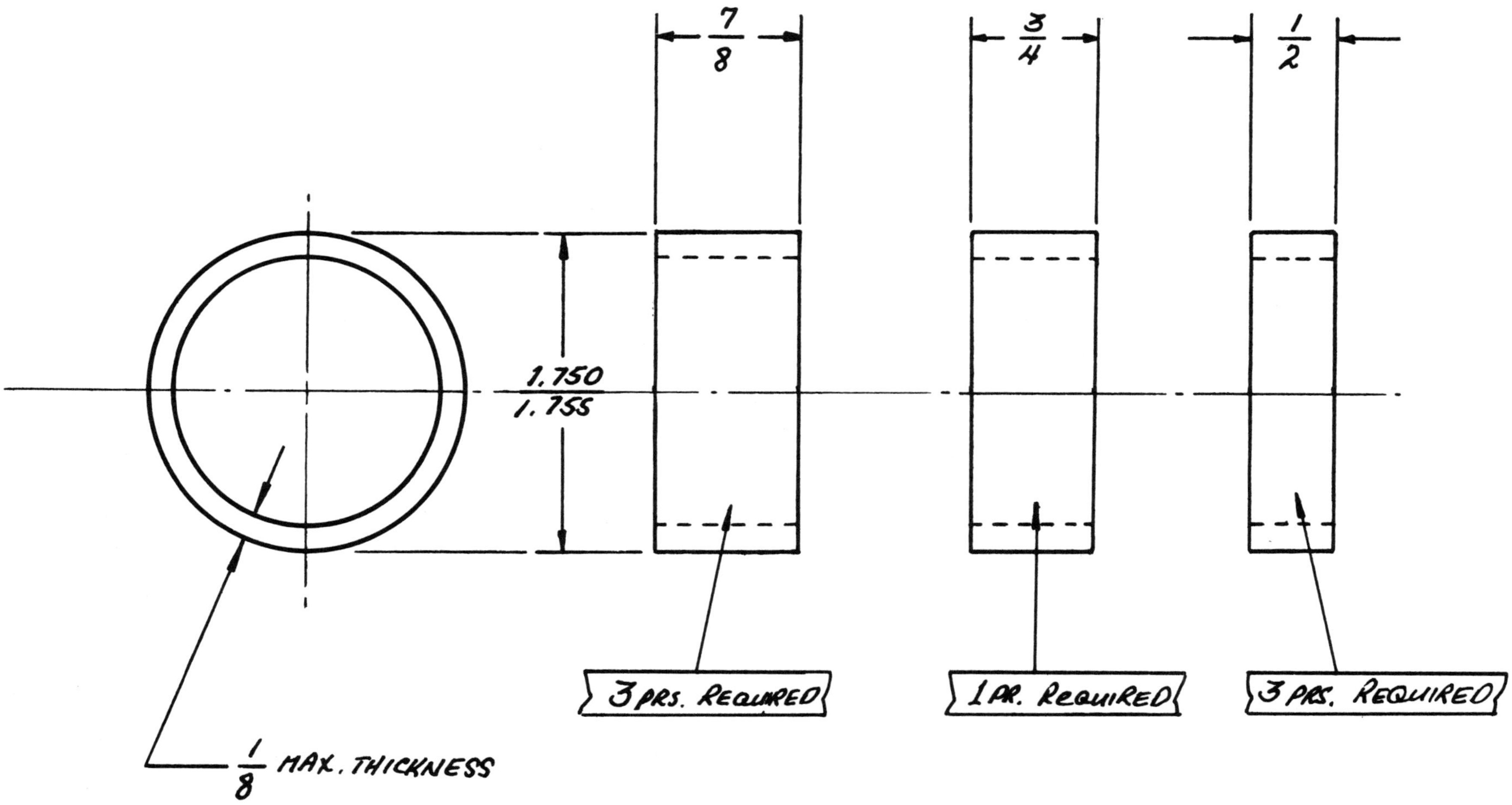

BAFFLE SPACERS

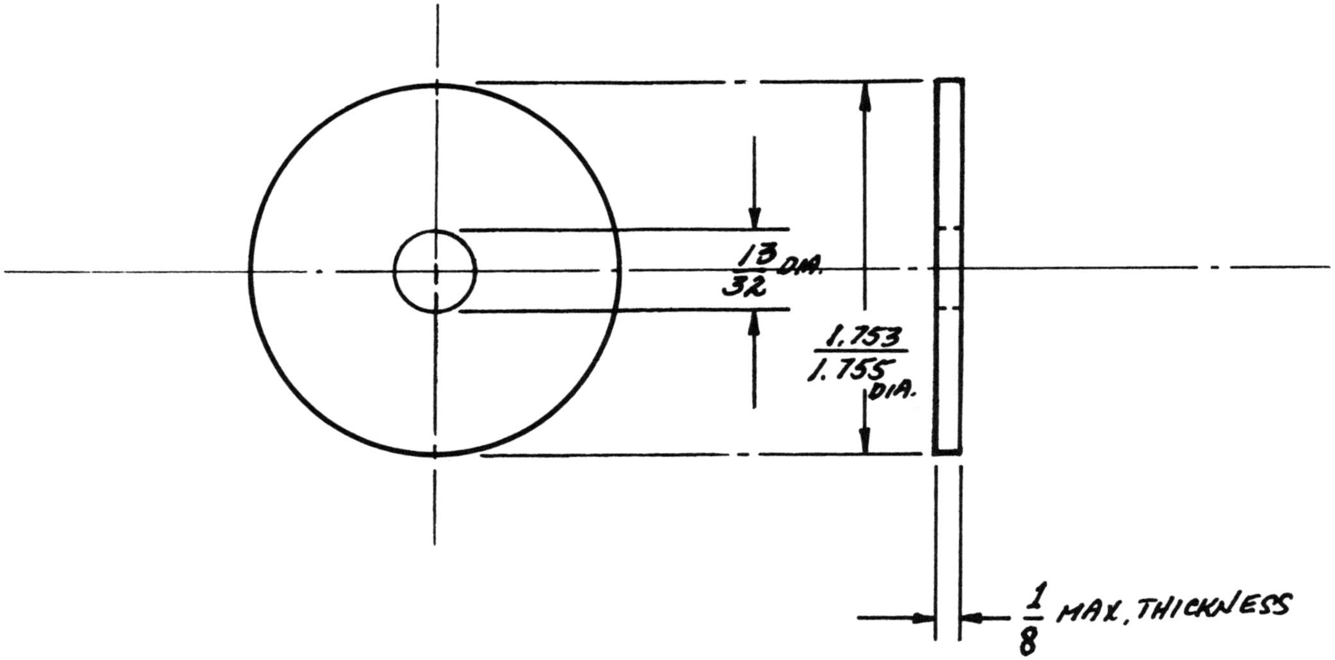

MATERIAL – STEEL SPACER DISK
"13-14 PRS. REQUIRED"

NOTE:
1. BAFFLE SUPPORT SHOULD FIT SNUGLY TO INSIDE DIAMETER OF SILENCER TUBE.
2. CENTER HOLE MUST BE PRECISELY IN LINE WITH BORE AXIS OF BARREL.

~BAFFLE SUPPORT~

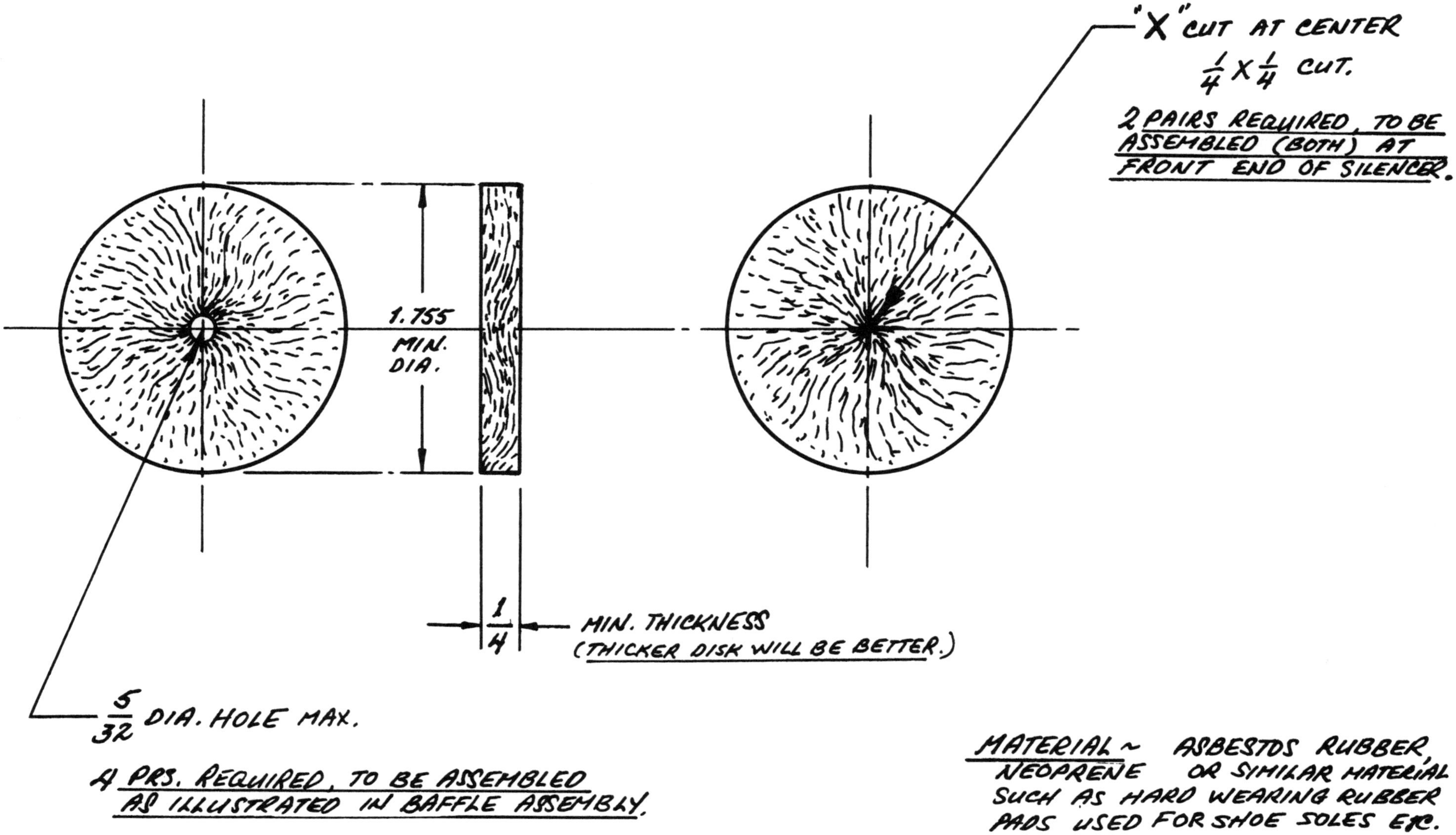
"X" CUT AT CENTER
1/4 X 1/4 CUT.
2 PAIRS REQUIRED, TO BE ASSEMBLED (BOTH) AT FRONT END OF SILENCER.
1.755 MIN. DIA.
1/4
MIN. THICKNESS (THICKER DISK WILL BE BETTER.)
5/32 DIA. HOLE MAX.
4 PRS. REQUIRED, TO BE ASSEMBLED AS ILLUSTRATED IN BAFFLE ASSEMBLY.
MATERIAL ~ ASBESTOS RUBBER, NEOPRENE OR SIMILAR MATERIAL SUCH AS HARD WEARING RUBBER PADS USED FOR SHOE SOLES ETC.

~ BAFFLE DISK ~

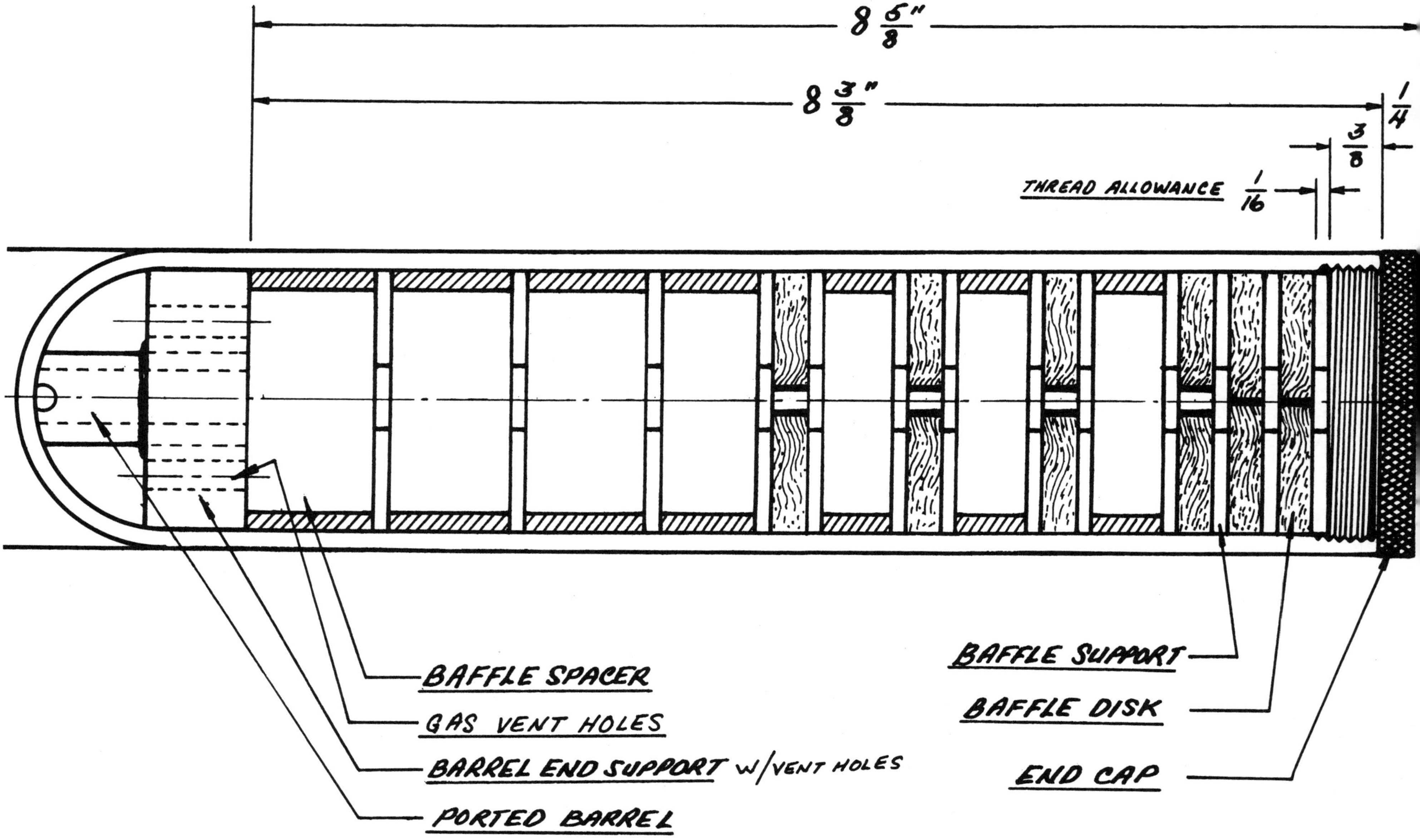

Detail ~ "BAFFLE ASSEMBLY"

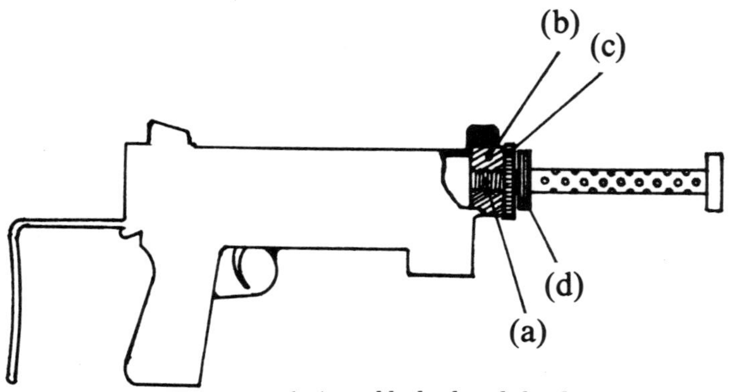

1. Assemble the threaded end (rear) of barrel (a) to receiver plug (b).
2. Tighten barrel nut (c) after assembly of barrel to receiver.

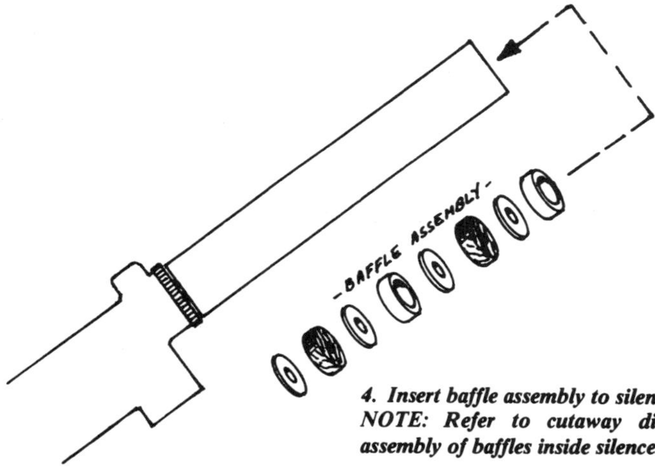

4. Insert baffle assembly to silencer tubing.
NOTE: Refer to cutaway diagram for proper assembly of baffles inside silencer tube.

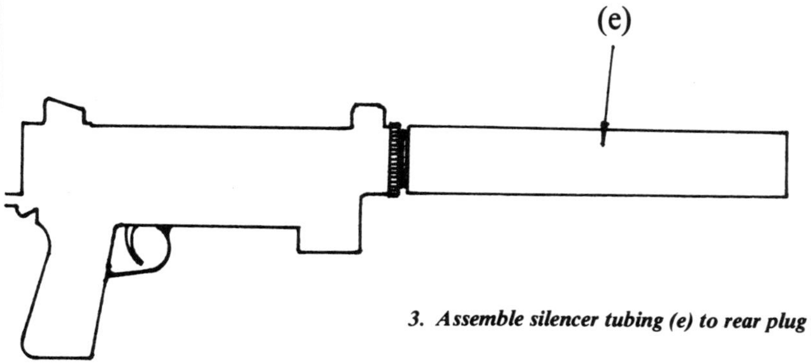

3. Assemble silencer tubing (e) to rear plug (d).

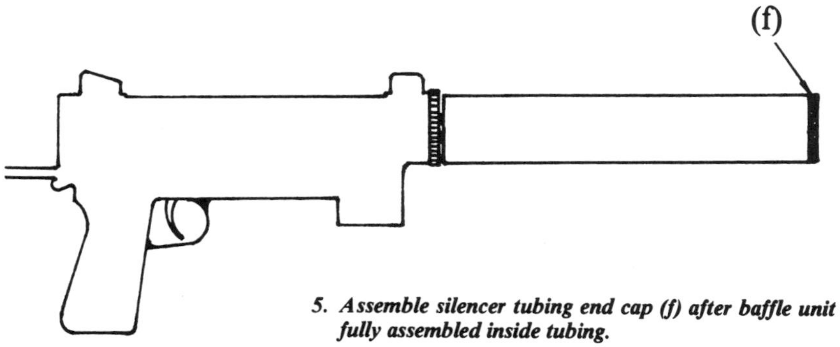

5. Assemble silencer tubing end cap (f) after baffle unit is fully assembled inside tubing.

V. SIMPLIFIED SMG SILENCER PLANS

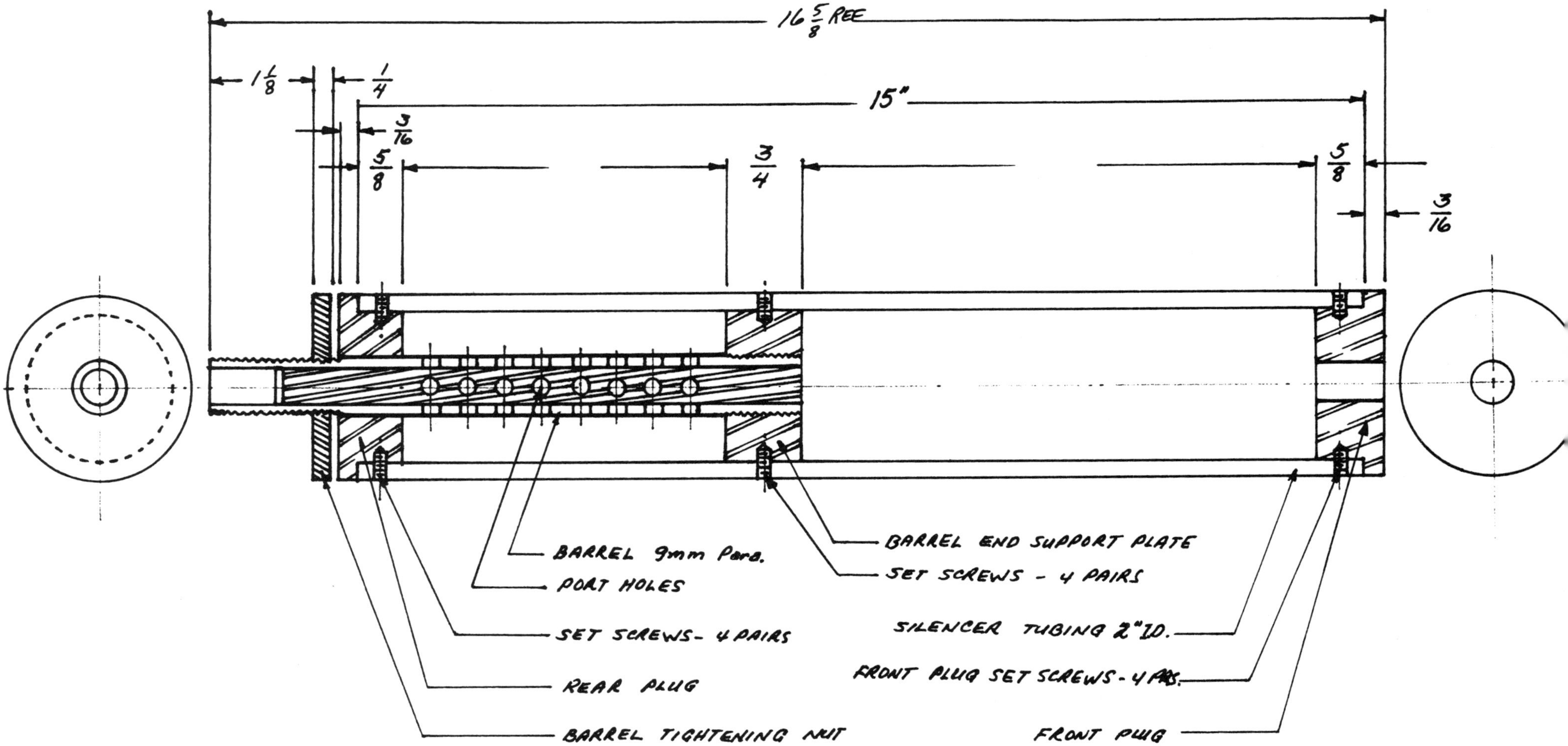

-S.M.G. SILENCER TUBING ASSEMBLY - SIMPLIFIED-

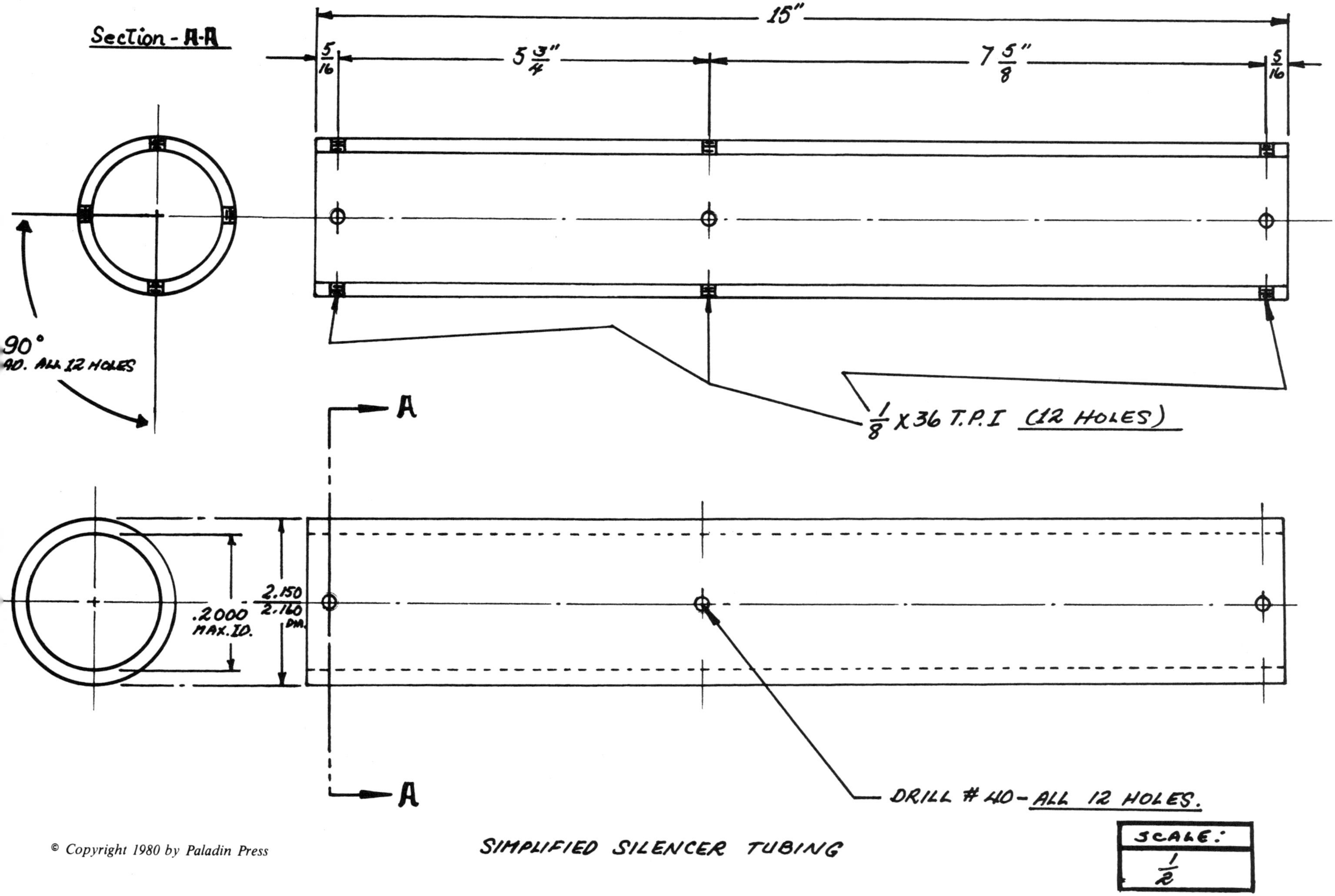

SIMPLIFIED SILENCER TUBING

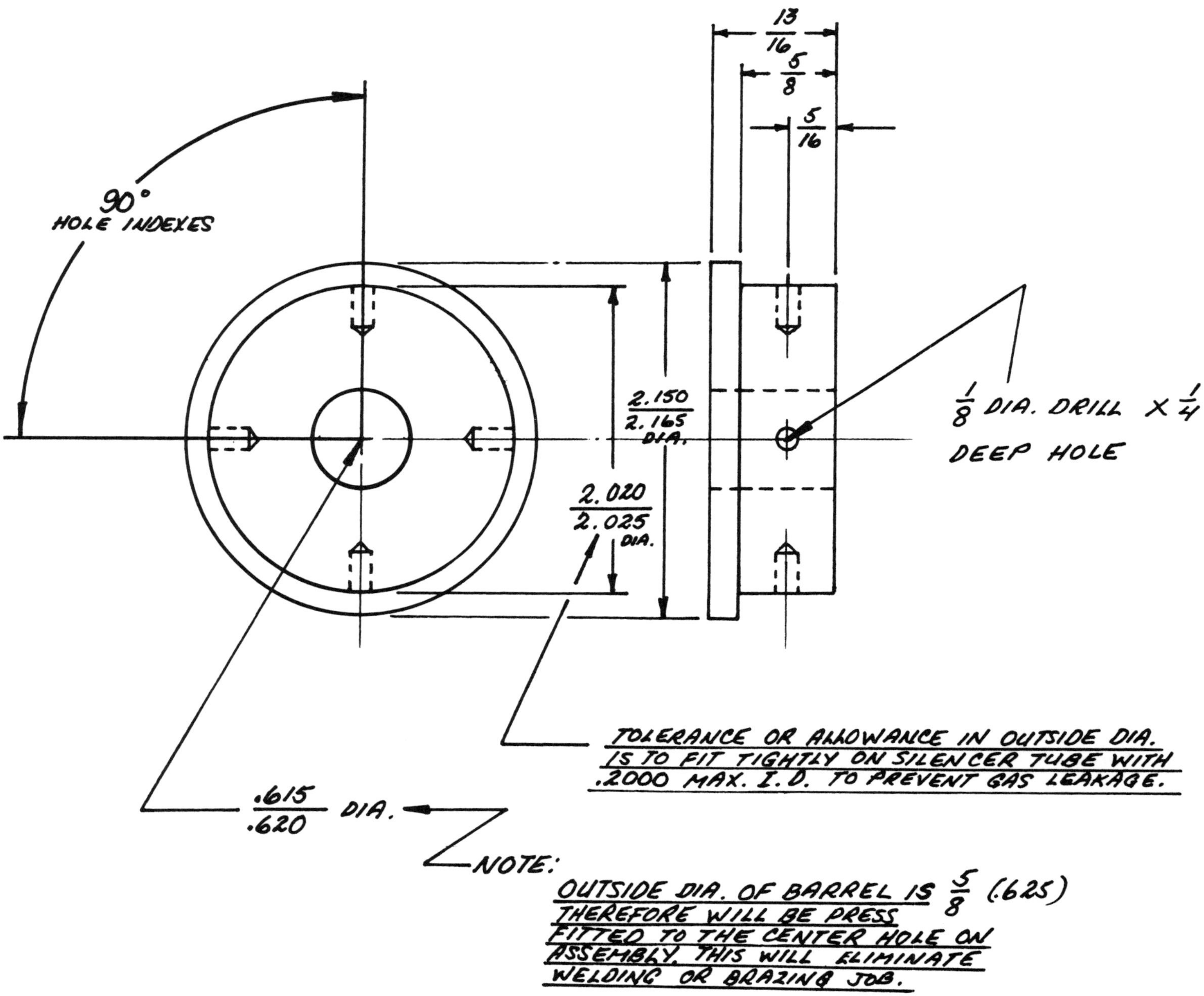

SILENCER TUBING-REAR PLUG

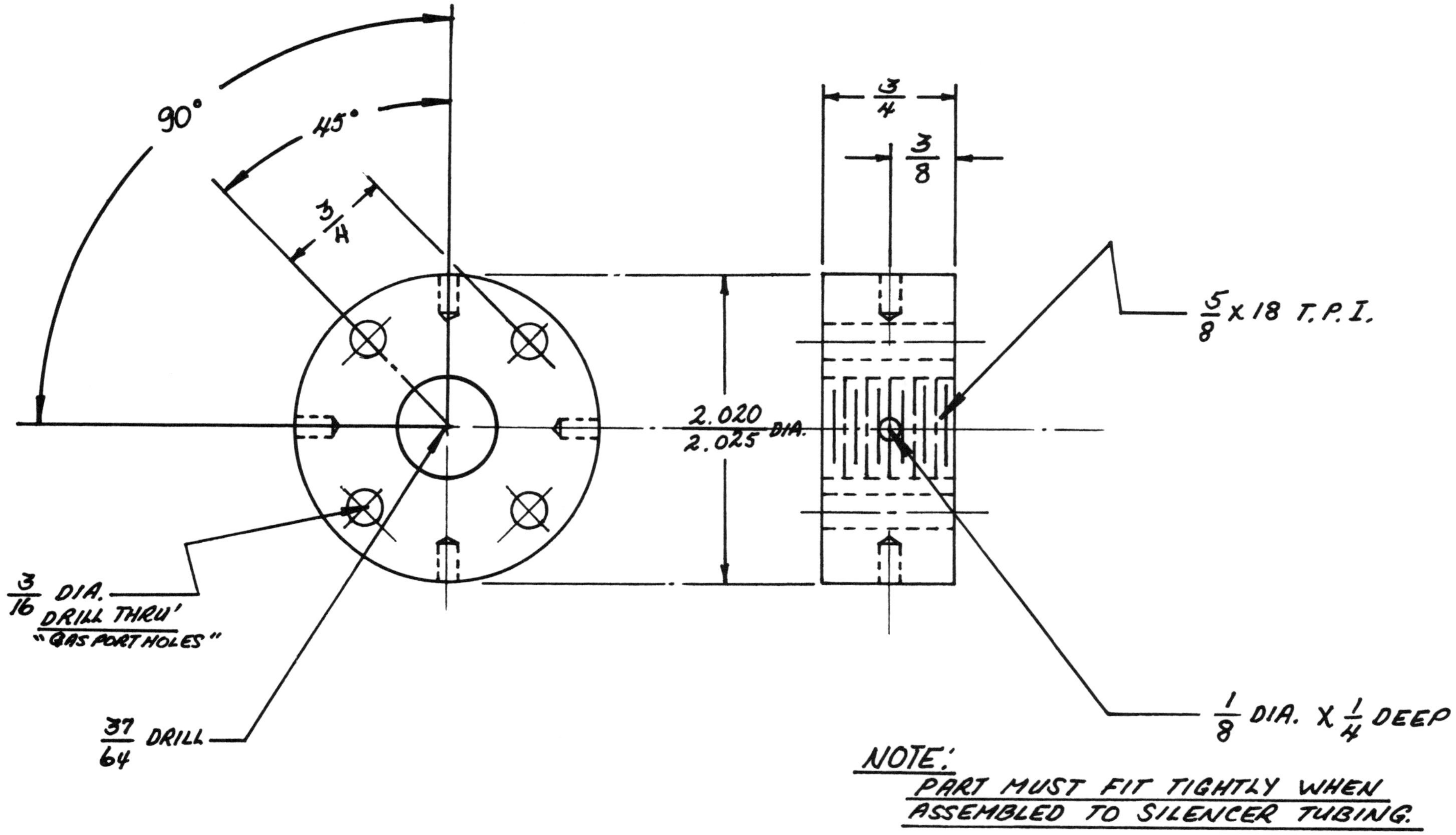

BARREL END SUPPORT

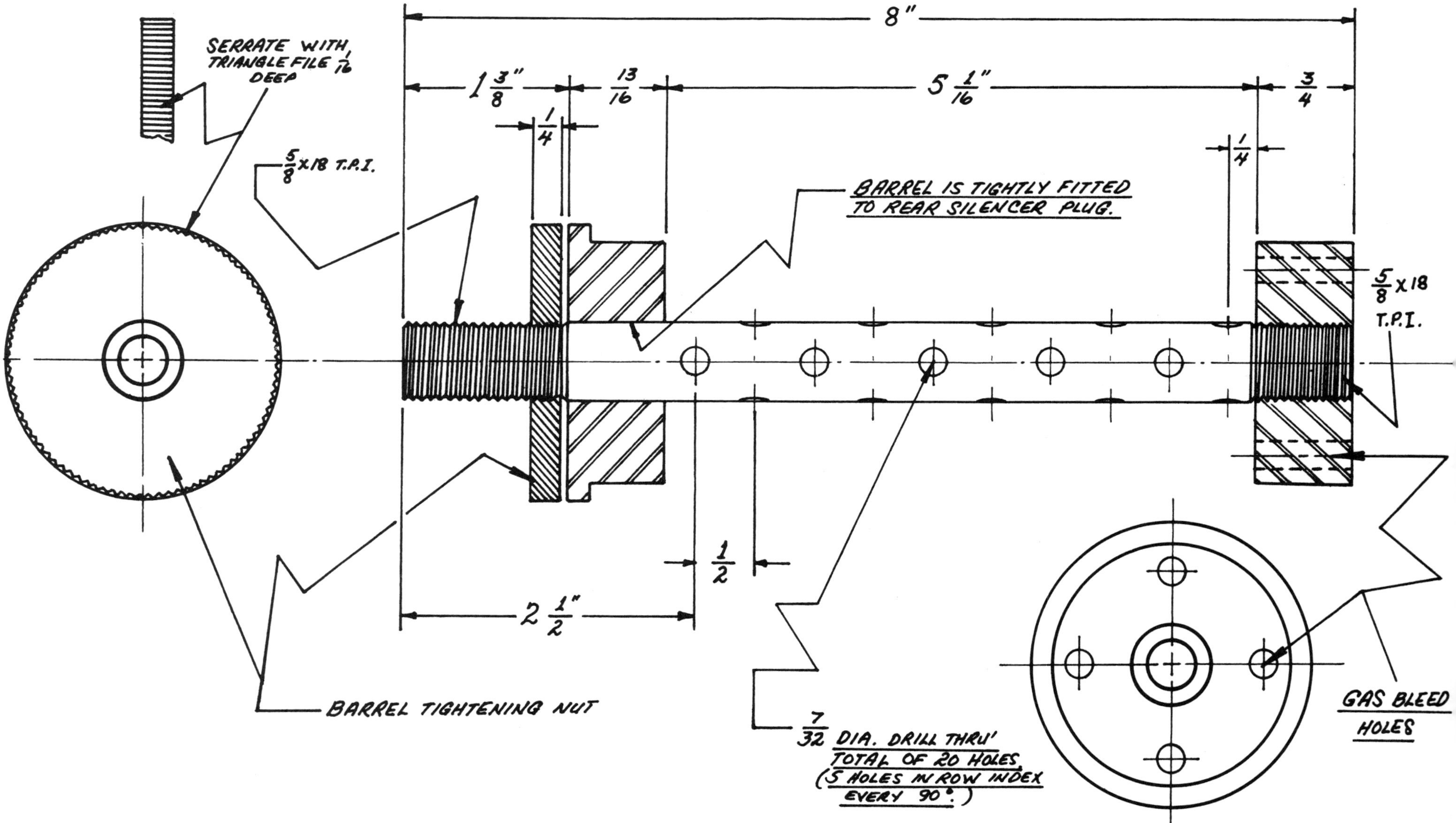

ASSEMBLY - BARREL UNIT

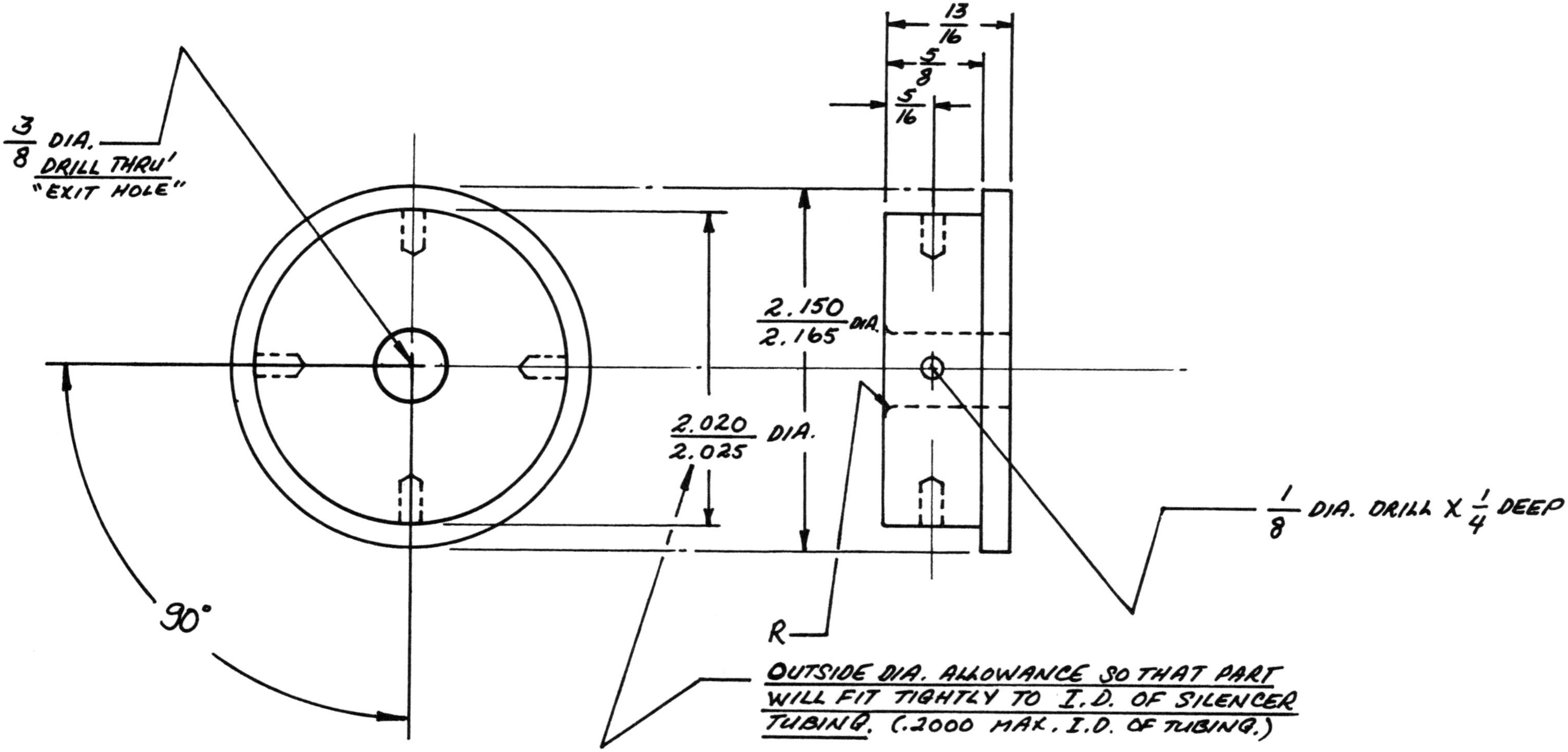

NOTES.

1. POLISH SMOOTH WITH FINE EMERY CLOTH ALL OUTSIDE SURFACE.
2. LAP EXIT HOLE BEFORE ASSEMBLY.

SILENCER TUBING-END PLUG

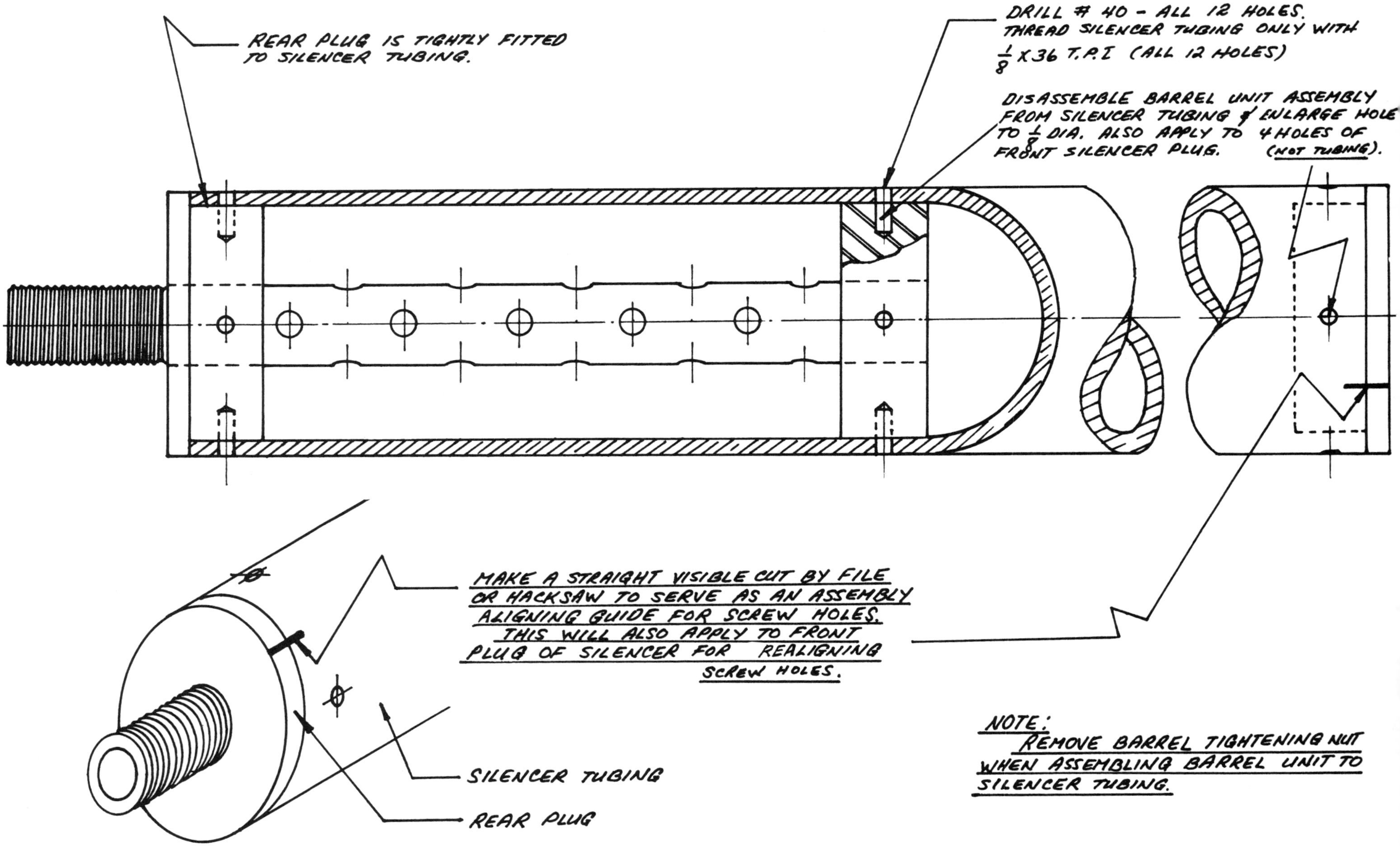

BARREL UNIT/SILENCER TUBING ASSY.

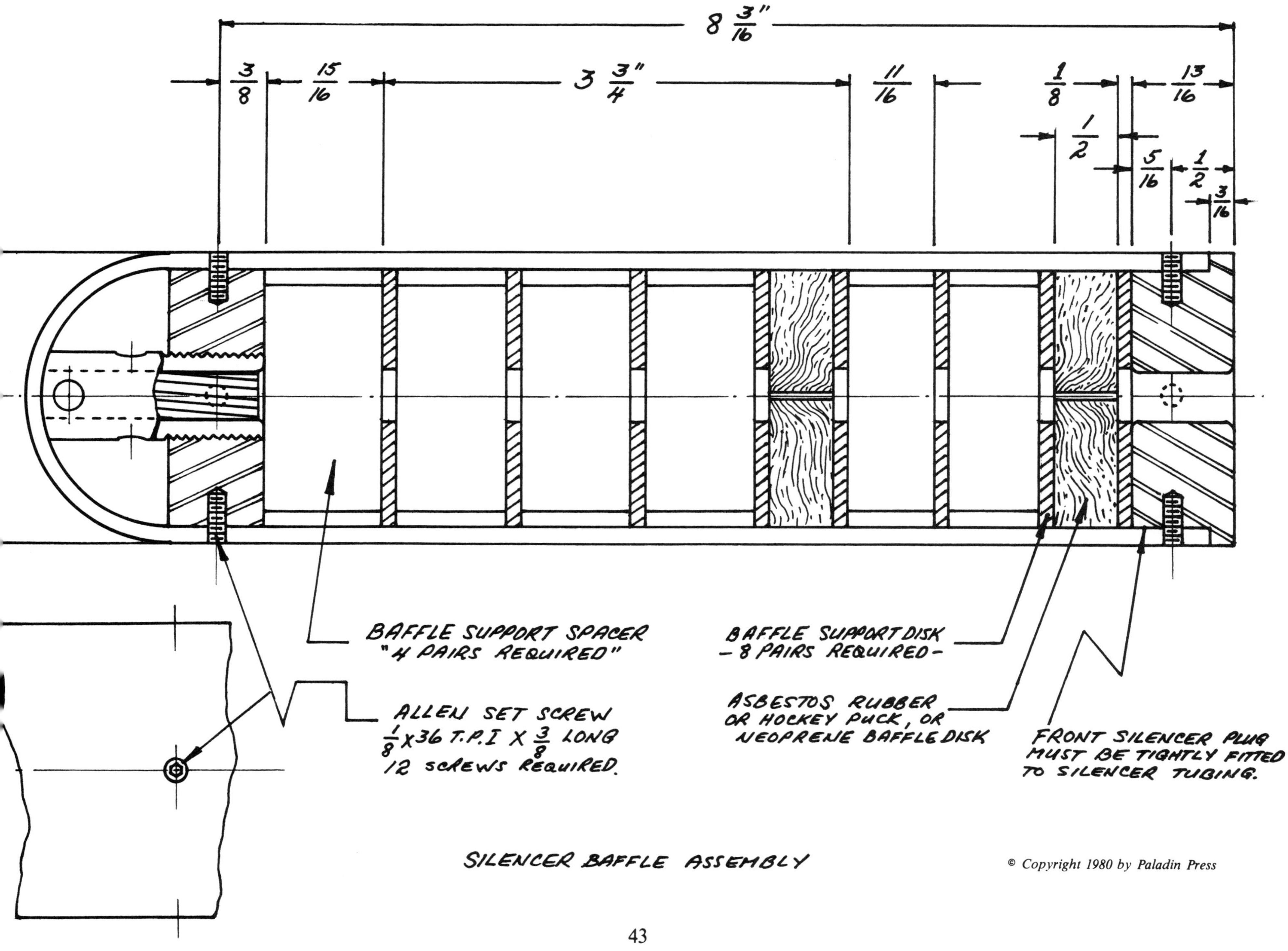

SILENCER BAFFLE ASSEMBLY

VI. PISTOL SILENCER — .32 & .380 ACP

The silencer to be discussed herewith is applicable to any pistol with the barrel enclosed inside the receiver tubing. To assemble a silencer for this type of barrel arrangement, the barrel tightening nut must be used as a base for the silencer tubing.

This is a good arrangement for a silenced pistol, since the barrel is not ported and bullet velocity is not affected. However, this system requires a subsonic bullet such as the cal. .32 or .380 ACP.

Applying this system to cal. .22 long rifle ammunition may not be practical unless a longer barrel, that can be ported, is utilized. The portion protruding from the receiver tube must be ported and fully enclosed inside the silencer tube as on the extension tube. The subsonic cal. .22 short, with a velocity similar to the .380 ACP cartridge, must be used.

Another pistol ammunition adaptable to this system is the cal. .25 ACP. Like the .32 ACP, it is well below sonic velocity, and is a very practical cartridge for this application. However, its stopping power cannot be compared to the .32 and .380.

This system can be applied to large numbers and varieties of small automatic pistols on the market. In the case of Ruger and Hi-Standard automatic pistols, the barrels are fully exposed and can be ported. Porting is necessary to reduce the velocity of the .22 LR ammunition to subsonic level. The baffle system described here is adaptable once the barrel has been ported. The full barrel length must be fully enclosed in the silencer tubing.

Silencer Tubing Construction

As previously described in the Smg silencer chapter, the first part to be built is the silencer tubing.

Tubing for the pistol silencer can be cut from readily available seamless tubing such as, aluminum, brass, or even water pipe.

The overall length of the silencer tubing should not be less than 8 inches. The inside diameter should be from .986 to 1.000 inch. It will have a minimum thickness of .125 inch for threading depth allowance.

In selecting the tubing material, inspect the inside and outside surface for cracks or holes through which pressure might escape, especially if water pipe is used. Make sure your tubing is straight. Crooked tubing may cause misalignment of the bullet leaving the extension tube.

After cutting the tubing to the specified length, grind and polish the surface ends smooth with a fine emery cloth. After it is polished, thread both ends with 1.173 x 7 TPI. Both ends have different lengths to be threaded. The rear end of the tubing will be threaded 3/8 inch in length and the other end 1/4 inch. In some cases longer threading is necessary for tightening allowances on both ends when the front and rear silencer plugs are assembled. After the threading operation is accomplished, the external surface of the tubing can be polished with rough emery cloth, then finished with fine cloth. The tubing is now ready for bluing.

Replacement Front Barrel Retainer & Rear Silencer Plug

After the silencer tubing, the next part to be built is the replacement front barrel retainer. It also acts as the rear silencer plug.

Due to the enclosed barrel arrangement of the pistol, a new type of barrel retainer must be built. It will double as a silencer tubing base.

The ideal material for this is round stock with an outside diameter of 1.350 to 1.355 inch.

First trim (lathe trim) the outer diameter on the end to be threaded down to 1.173 to a length of 5/16 inch. If a lathe is available, three operations can be accomplished; drilling the center hole, trimming the end to be threaded, and trimming the other end to which it is inserted onto the receiver body. After all trimming is done, knurling or threading is easily accomplished. Most proficient machinist have their own ways of doing this.

If you are working with hand tools, first trim the end to be threaded with a sharp flat file, holding the part in a vise. Filing must be done uniformly. Do not file too deeply on one side and too shallowly on the other. Keep your filing strokes uniform until the perfect dimension is reached. This will also apply to the portion that is inserted to the receiver body.

After trimming, locate the center of the round stock and mark it with a center punch. Do not drill it yet unless you have a lathe or good drill press with a built-in vise for precision. Grip the other end of the round stock, exposing the area to be threaded. Oil will expedite the threading operation. It will also prevent tool breakage for the careful machinist. After threading, remove the part from the vise and cut it to the specified length.

Cutting can be done on a lathe or by a hacksaw. When cutting by hacksaw, the surface is not usually straight. Therefore, it is necessary to file off the excess uniformly, polishing it with emery cloth.

Now put the part in the vise, threaded area facing down. Make sure the vise jaws are padded with either wood or rubber to prevent denting of the threads. Locate the center of the part through which a hole is to be drilled. Always start with a smaller drill and gradually work up, finishing with the 27/64 drill. After the final hole size is reached, thread the hole with 9/16 x 18 TPI, to match the barrel end thread.

Remove the part from the vise and grip the part horizontally. The jaws now grip both ends of the part. It is now ready for checkering by hand. Use a small triangle file (not needle file) and file in any pattern you want. Turn the part each time one side is finished until the entire surface is checkered.

After knurling is completed, remove the part from the vise and break any external edges with emery cloth. The part must be refitted to the silencer tubing and barrel. Observe the fitting and make necessary adjustments.

Once fitting is accomplished, buff the part and put it aside for bluing when all the parts are completed. Continue to the next part.

Barrel Extension Tubing

As discussed earlier, the barrel is fully enclosed inside the receiver. It is therefore necessary to make an extension of the barrel for porting. The new replacement barrel is longer in front than the original. This is necessary so that the extended length of the retainer will serve as a base for the barrel extension tubing and silencer tubing.

First locate seamless tubing with an inside diameter of 3/8 inch and if possible, an outside diameter of 9/16 inch. If you have a lathe, build it by drilling a 3/8 inch hole in 9/16 inch round stock similar to that used in building the barrel. Rifling can be added with this method. However, since the bullet is loose inside the extension tubing bore (when using seamless tubing), a simple seamless tubing of steel, aluminum or brass will do as well. This will save you a lot of trouble over manufacturing one from solid stock.

The overall length required for the extension tube is 4 inches. The other end must be threaded with 9/16 x 18 TPI, the same thread used to thread the barrel end. This threaded end will match the barrel retainer nut. The threaded length is about 5/16 inch maximum. The unthreaded end of the tubing will be crowned in the same manner as the barrel. A small countersink or crown is drilled inside the rear threaded end of the tubing. This will prevent the bullet from shaving lead while leaving the muzzle end of the barrel.

After threading and crowning the specified portion of the part, the inside bore of the tubing must be lapped smooth like the bore of the barrel. This will help the bullet pass through smoothly, without snagging.

Note that the bullet must not be excessively loose. If so, the bore of the extension tubing is too big. The bore diameter of the extension tubing must be exactly the same as the diameter of the bullet, or slightly smaller without being overly tight. This prevents misalignment of the bullet's flight through the exit hole of the baffle assembly unit. It will also prevent excessive escape of gases around the bullet while inside the tubing, diverting more gases into the expansion chamber area through the vent or port holes of the tubing.

The port holes of the extension tubing are 1/8 inch or larger in diameter. There will be 6 holes in a row, graduated every 90 degrees, for a total of 24 holes. For an accurate description of port dimension and distances, refer to the diagram of barrel extension tubing.

After drilling port holes in the tubing, vigorously scrub the inside of the bore with a brass brush. Clean it with solvent degreaser, then relap with fine emery cloth or any other lapping materials.

Extension Tubing End Support

After completing the barrel extension tubing, we make the end support for it. The support plate will line up the end of the extension tubing with the baffle assembly exit holes and end cap exit hole. It will also act as a circulatory gas block for the expansion chamber enclosing the extension tubing.

First prepare round stock with an outside diameter of .986 to 1.000 inch. It must match precisely the inside diameter of the silencer tubing. The support plate will fit snugly inside the silencer tubing to prevent any movement at the end of the extension tube during battery. Cut a 3/8 inch length from the round stock bar. The cut surface must be ground evenly, then polished smooth with fine emery cloth.

After grinding and polishing, locate the center axis by using a compass or similar tool, and mark it with a center punch. Drill a hole on center with a 35/64 inch drill. For best results, start with small drill sizes and proceed with larger ones, until reaching the finished size. Drill a 60 degree countersink in the other end of the hole. This countersink is to be filled flush later with weld or brazing.

Note that the diameter of the hole, once press fitted together, is smaller than the outer diameter of the barrel extension tubing. In this case a slight trimming might be needed on the muzzle end of the extension tubing until the part fits snugly to the support plate hole or until it can be press fitted easily. Trimming the outer diameter of the extension tubing can be done on a lathe or by simply filing its end portion uniformly. DO NOT OVERFILE.

Once both the extension tubing and support plate are press fitted together, the countersunk portion is filled with weld or brazing. Be careful not to apply excessive heat. This will cause warping if the tubing is thin or of unknown steel. Another way of securing both parts is by threading them together to match.

After welding or brazing, re-polish all external surfaces. You can lathe-turn or file off excessive weld around the tubing. Check the bore diameter to make sure the heat has not affected its normal diameter. Redrill to specified dimension and straighten it up if it has been affected by the heat.

We now drill the vent holes of the support plate. These holes will circulate gases escaping from the expansion chamber to baffle channels in front of it until it diminishes slowly into the exit holes of the rubber plug. The remaining pressure of the gases left behind by the bullet will force its way out to the baffle channels and exit holes of the rubber baffles.

The minimum number of holes that should be drilled through the plate is 4 to 8. The minimum diameter of the holes is 3/32 to 1/8 inch, depending on the size of your plate. There is no specific graduation of hole placement as long as they are drilled uniformly around the plate and positioned so that the baffle spacer is not blocking or covering the hole, preventing free venting of gas.

Silencer Tubing End Cap

After perfect fitting of the barrel unit and silencer tubing is achieved, we make the silencer tubing end cap. This part seals the baffle unit assembly inside the silencer tube.

Obtain round stock of 1.350 OD, which is the same outside diameter as the silencer tubing. A slightly larger stock is acceptable. Lathe-turn the end and reduce the diameter to 1.173 inch, with a length of 3/16 inch. This will be threaded with 1.173 x 7 TPI. Thread it on the lathe or by hand. If the part is threaded on a lathe, it can also be drilled at the center with a 3/8 inch diameter hole through to a depth of 1/2 inch while it is still chucked. Start with a smaller size, drilling up to the final size.

Proceed knurling it on the lathe. Knurl before or after threading, as you prefer. After knurling, cut the part to the specified length of 3/8 inch and grind the cut surface. Finish by polishing with emery cloth. Sharp edges and surfaces must also be polished and the inside diameter of the center hole must be lapped.

Assemble the part on the silencer end and check for final fitting. Prepare the part for bluing once it is fitted properly.

Baffle Spacers

We are now ready to build the baffle unit assembly. First calculate the distance from the extension tubing end surface to the rear of the silencer end cap when assembled to silencer. This distance is your space for the baffle unit assembly.

The cutting of baffle spacers will depend on the number and thickness of rubber baffles used. Thick rubber baffles require shorter spacers.

It is therefore up to you to calculate the exact lengths of your baffle, depending on the material you have chosen.

The baffle spacers must have an outer diameter that causes them to be slightly loose when assembled inside the silencer tubing. Normal thickness will be .100 for this particular type. Its thickness must not block the vent holes of the extension tubing support plate.

Materials that can be cut for spacers range from seamless aluminum tubing, to brass tubing, to even water pipe.

These materials are readily available from any hardward or building supply store.

Baffle Supports

The baffle support disks used here are usually found at hardware and plumbing stores. They come in a variety of sizes and types of materials such as steel, bronze, brass, and aluminum.

Baffle supports for the pistol silencer require an outside diameter of .986 to 1.000 inch, with a minimum thickness of 1/16 inch. The number required will depend on the length and thickness of the rubber baffle and also the length of the spacers.

The center hole of the baffle support disk is about 3/8 inch or just large enough to clear the bullet as it exits through the baffle channels.

Rubber Baffle Disk

The last materials required to complete the silencer unit are the rubber baffles. These seal the gases inside the silencer tubing after the bullet has passed through the last rubber baffle disk near the exit hole of the silencer end cap.

The baffle must fit slightly tight when assembled inside the silencer to seal off the portion of gases between the pockets of the spacers. A small hole can be drilled through the baffle, normally 1/16 to 1/8 inch for cal. .25 to .380 ACP. You can even leave the baffle blank if you wish, without drilling center holes. Then it's up to the bullet to make its own hole when exiting.

When selecting your baffle materials, try to find a good, thick (1/4 to 1/2 inch) hard-wearing rubber material. Some good materials: neoprene pads, shoe soles, and hockey pucks. Of course the best to use, if you can find it, is asbestos rubber used by most manufacturers in their silenced weapons. This material is not affected by the heat of the powder charge during battery.

Refer to the cutaway diagram on assembling the baffle assembly unit inside the silencer tube.

Also, a separate chapter illustrates how to build an improvised baffle cutter.

Note: Make sure all rubber baffle disks are supported on both sides by a baffle support disk and are compressed as tightly as possible when fully assembled. This prevents the bullet from pushing the disks out of the assembly.

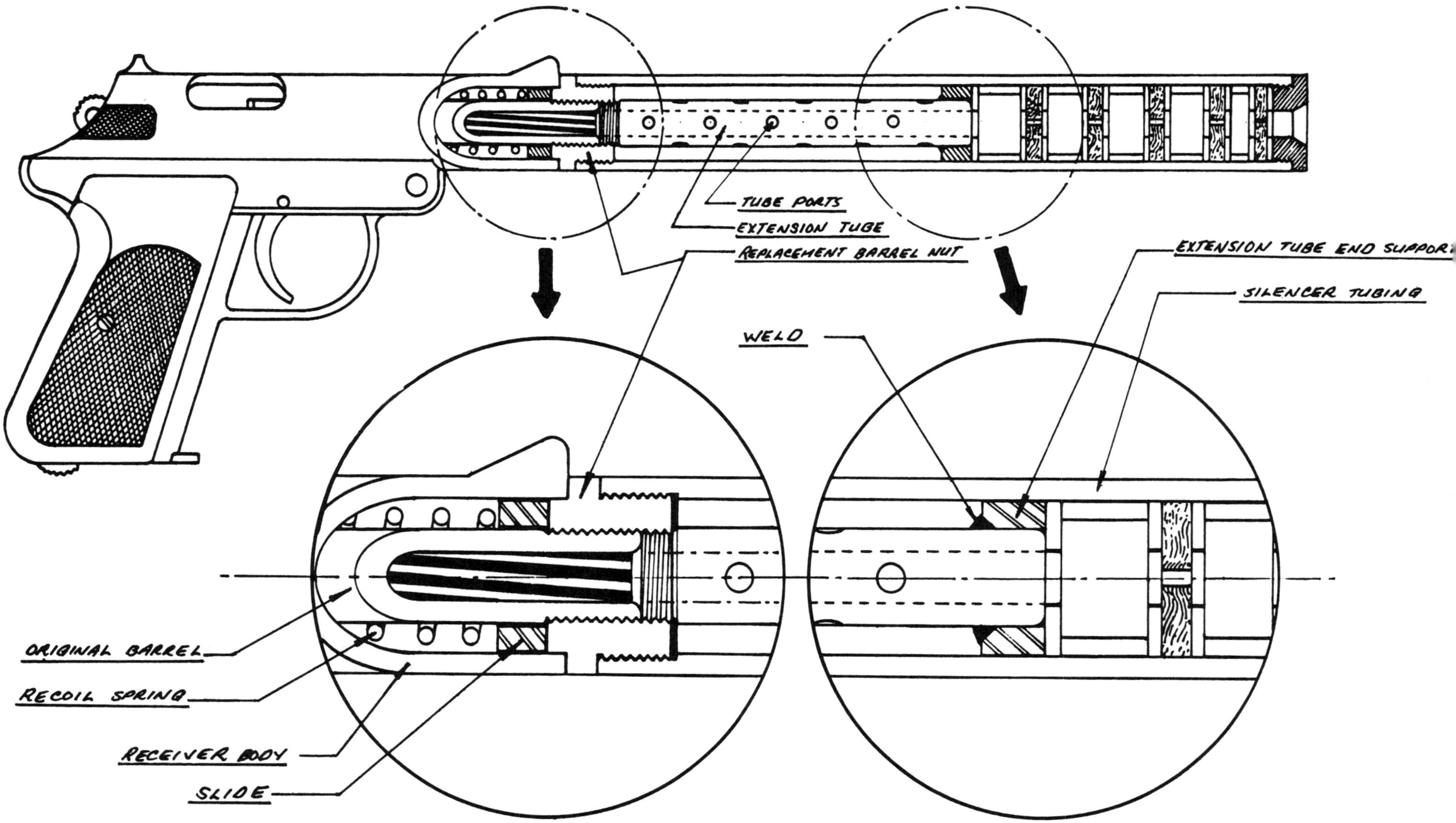

SECTION DETAIL - SILENCER ASSEM.

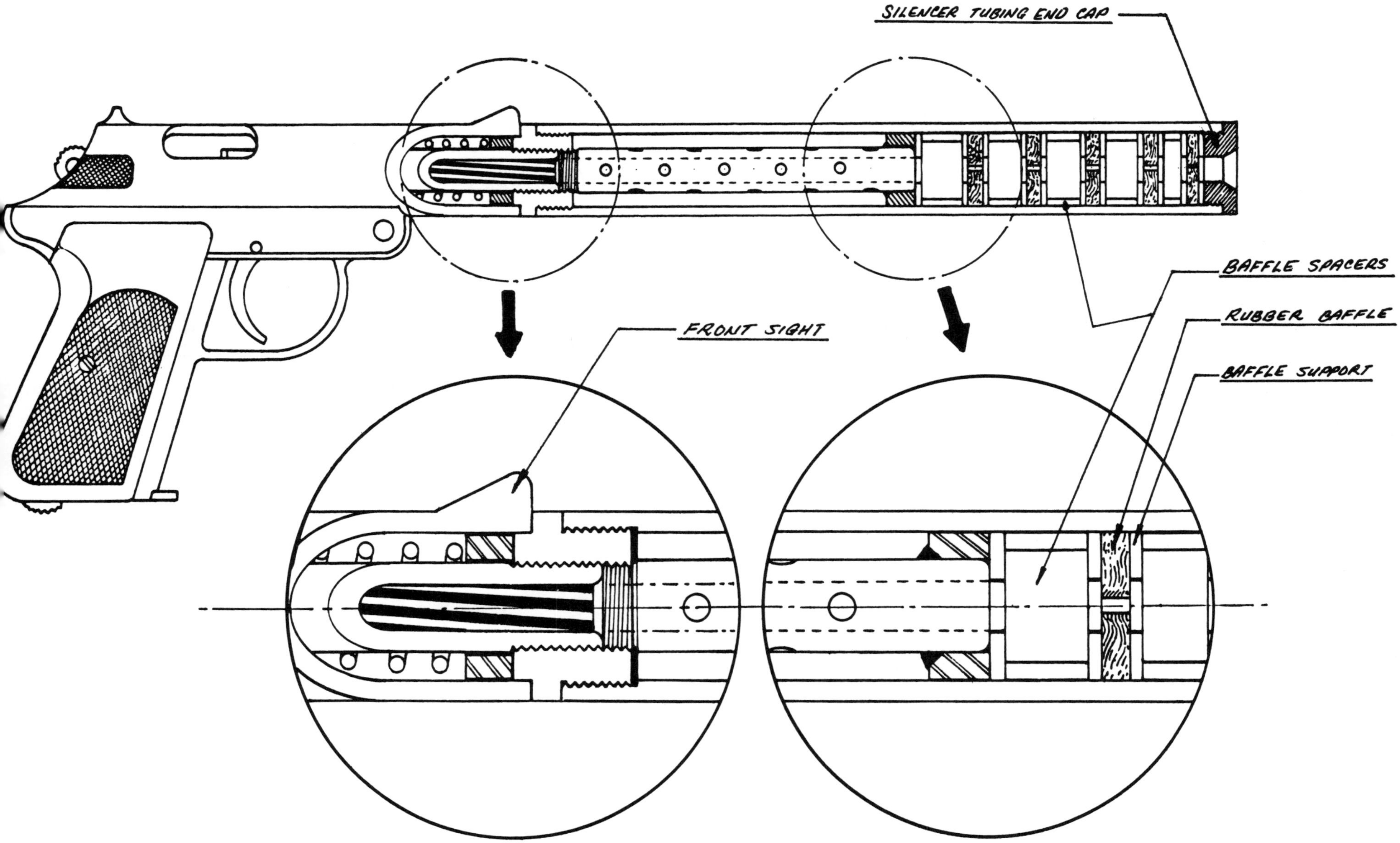

SECTION DETAIL - SILENCER ASSEM.

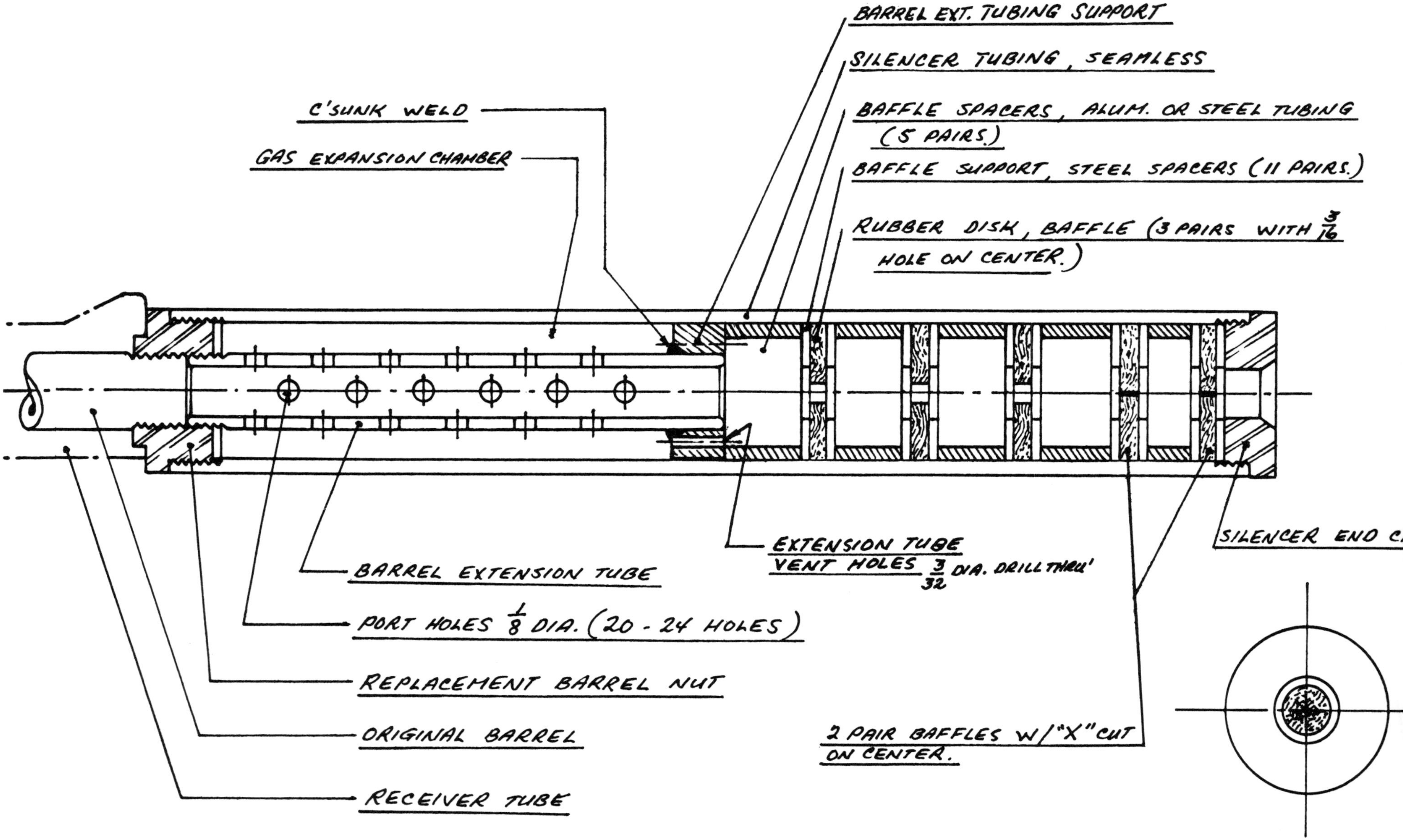

SILENCER DETAIL ~ Cal. 32 & .380 A.C.P.

SCALE
Full

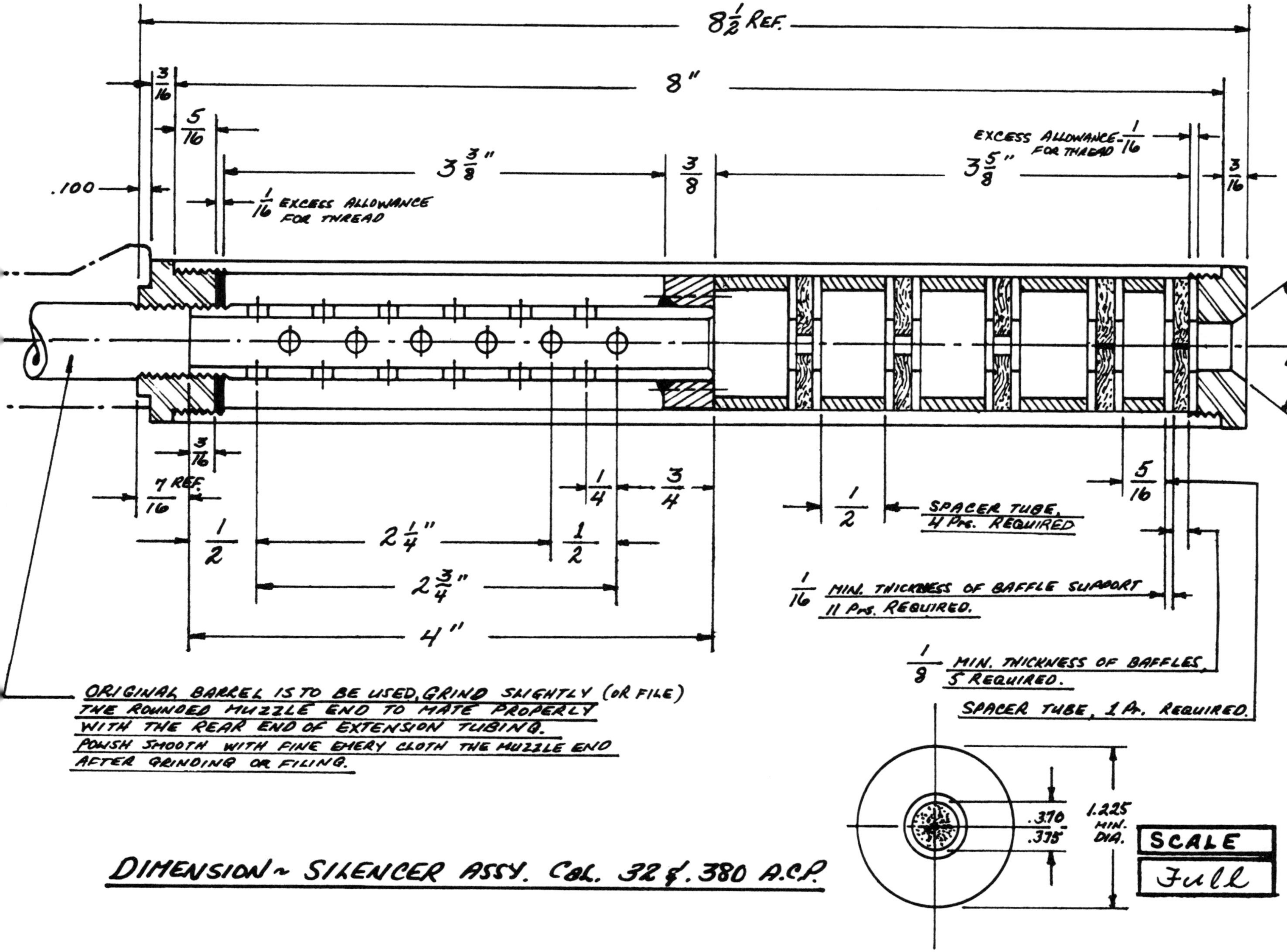

8 1/2 REF.
8"
3/16
5/16
.100
1/16 EXCESS ALLOWANCE FOR THREAD
3 3/8"
3/8
3 5/8"
EXCESS ALLOWANCE FOR THREAD 1/16
3/16
60° C'SUNK
3/16
7/16 REF.
1/4
3/4
1/2
2 1/4"
1/2
2 3/4"
4"
1/2
SPACER TUBE, 4 PCS. REQUIRED
5/16
1/16 MIN. THICKNESS OF BAFFLE SUPPORT 11 PCS. REQUIRED.
1/8 MIN. THICKNESS OF BAFFLES, 5 REQUIRED.
SPACER TUBE, 1 PC. REQUIRED.
ORIGINAL BARREL IS TO BE USED, GRIND SLIGHTLY (OR FILE) THE ROUNDED MUZZLE END TO MATE PROPERLY WITH THE REAR END OF EXTENSION TUBING.
POLISH SMOOTH WITH FINE EMERY CLOTH THE MUZZLE END AFTER GRINDING OR FILING.
.370
.375
1.225 MIN. DIA.
SCALE
Full
DIMENSION ~ SILENCER ASSY. CAL. 32 & .380 A.C.P.

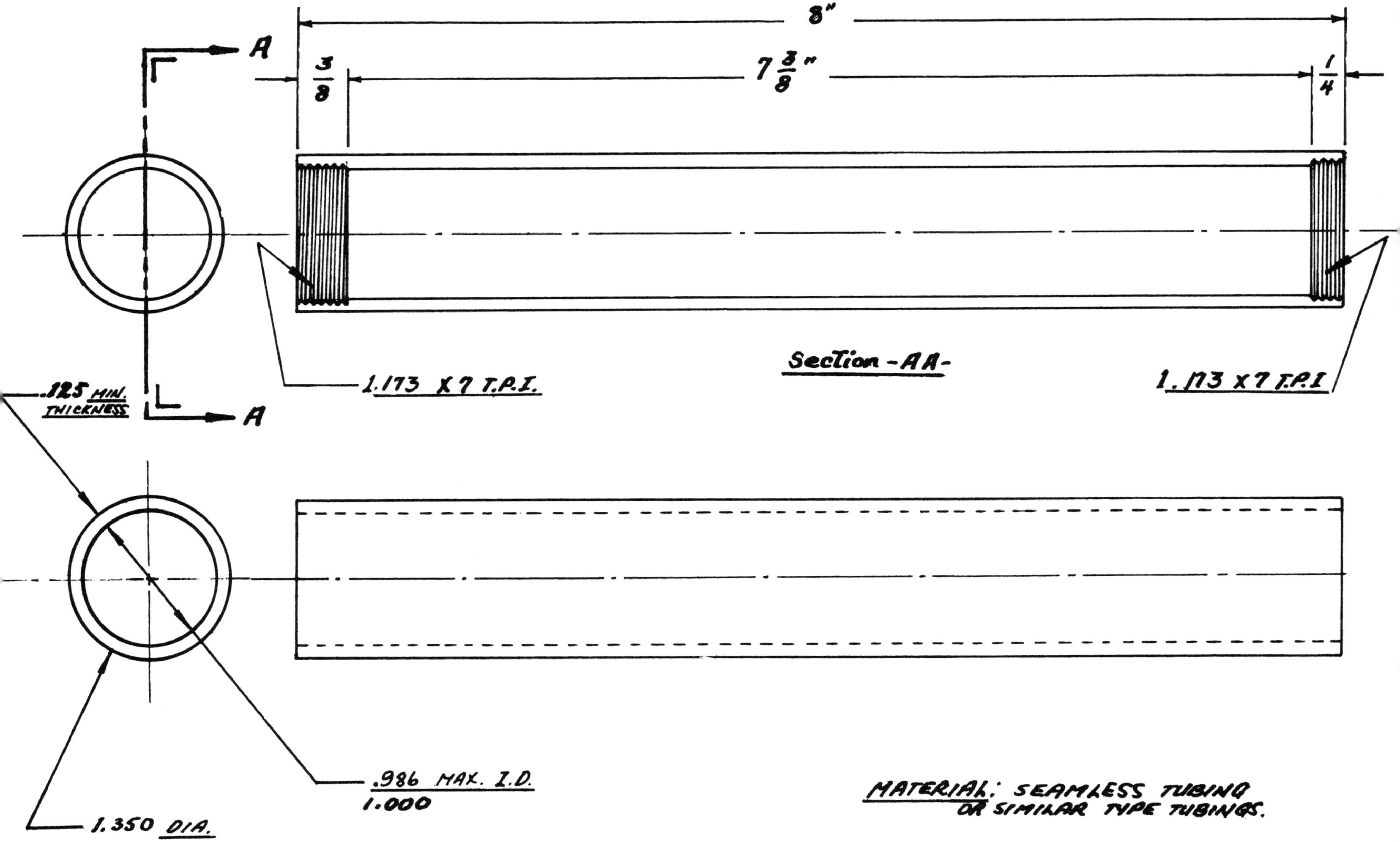

SILENCER TUBING DETAIL

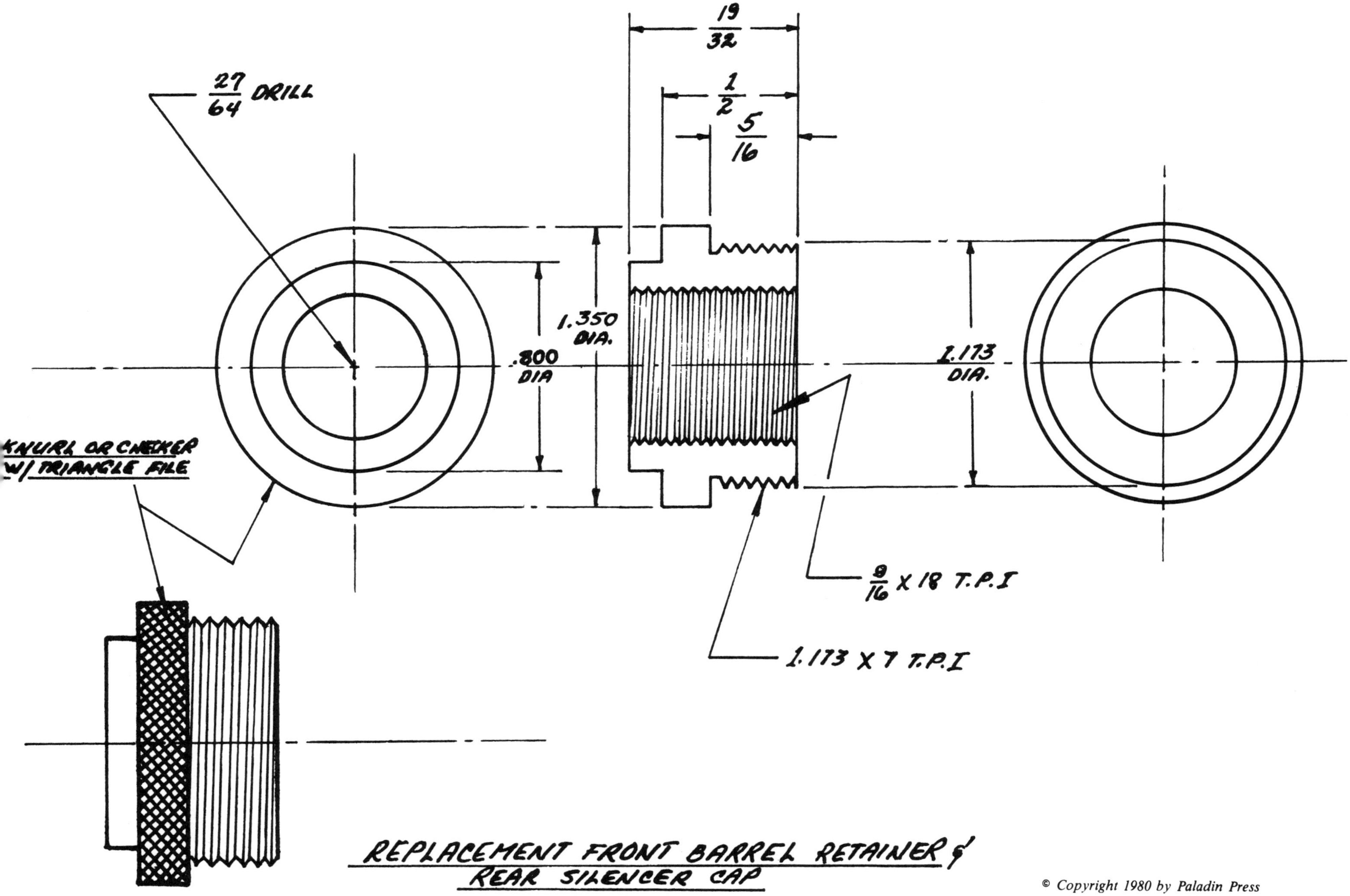

REPLACEMENT FRONT BARREL RETAINER & REAR SILENCER CAP

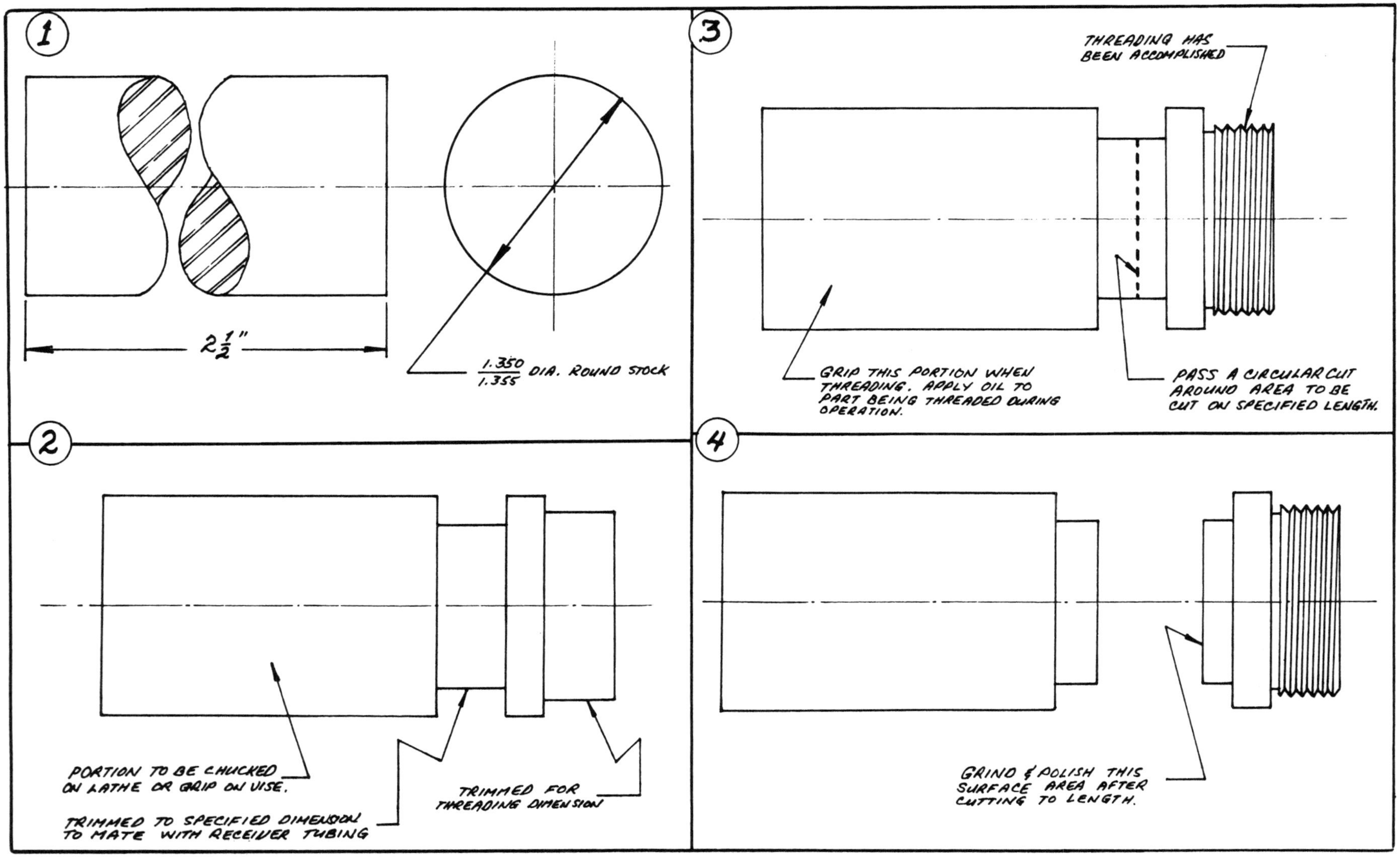

1
2 1/2"
1.350 / 1.355 DIA. ROUND STOCK
3
THREADING HAS BEEN ACCOMPLISHED
GRIP THIS PORTION WHEN THREADING. APPLY OIL TO PART BEING THREADED DURING OPERATION.
PASS A CIRCULAR CUT AROUND AREA TO BE CUT ON SPECIFIED LENGTH.
2
PORTION TO BE CHUCKED ON LATHE OR GRIP ON VISE.
TRIMMED TO SPECIFIED DIMENSION TO MATE WITH RECEIVER TUBING
TRIMMED FOR THREADING DIMENSION
4
GRIND & POLISH THIS SURFACE AREA AFTER CUTTING TO LENGTH.

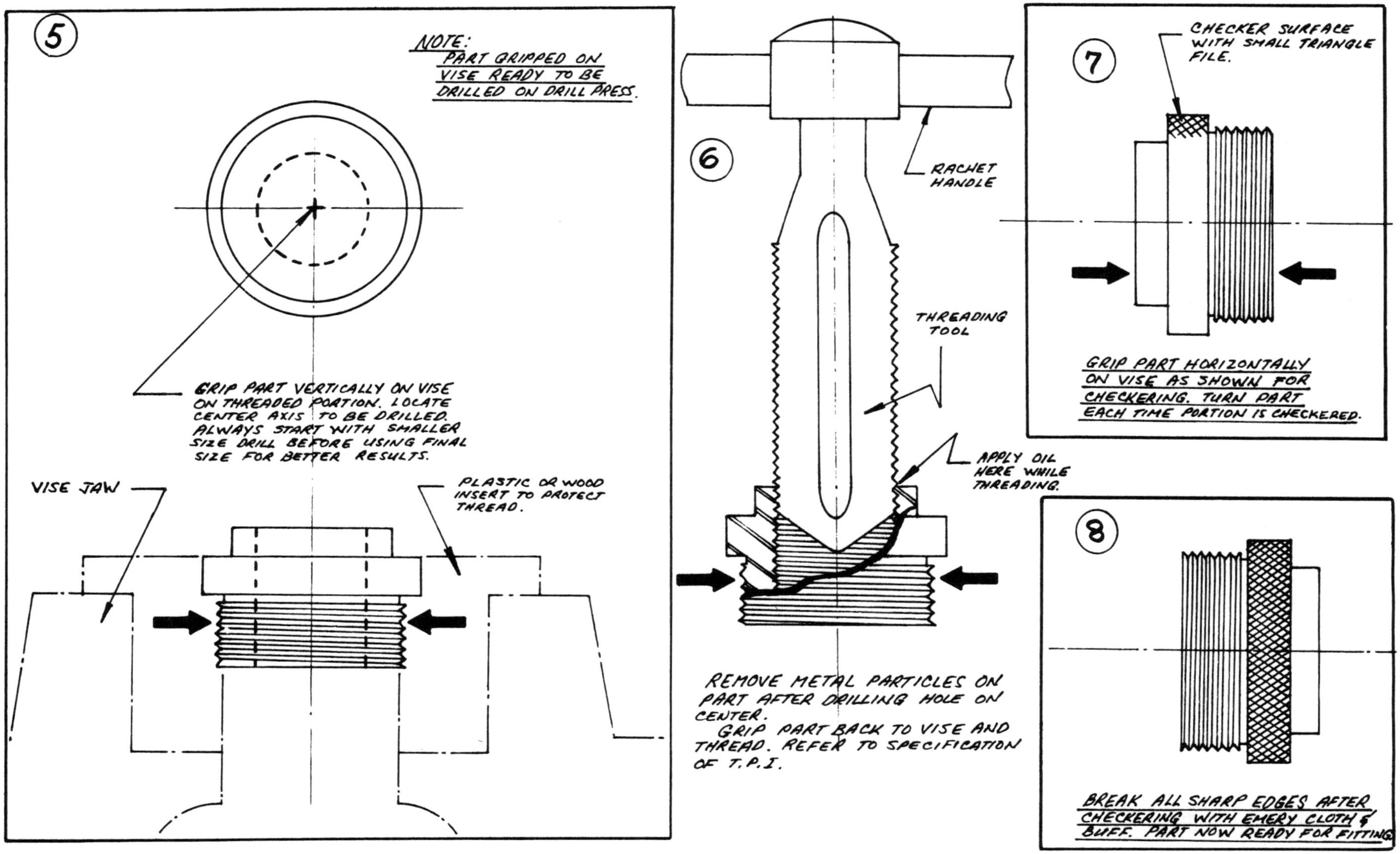
5
NOTE:
PART GRIPPED ON VISE READY TO BE DRILLED ON DRILL PRESS.
GRIP PART VERTICALLY ON VISE ON THREADED PORTION. LOCATE CENTER AXIS TO BE DRILLED. ALWAYS START WITH SMALLER SIZE DRILL BEFORE USING FINAL SIZE FOR BETTER RESULTS.
VISE JAW
PLASTIC OR WOOD INSERT TO PROTECT THREAD.
6
RACHET HANDLE
THREADING TOOL
APPLY OIL HERE WHILE THREADING.
REMOVE METAL PARTICLES ON PART AFTER DRILLING HOLE ON CENTER.
GRIP PART BACK TO VISE AND THREAD. REFER TO SPECIFICATION OF T.P.I.
7
CHECKER SURFACE WITH SMALL TRIANGLE FILE.
GRIP PART HORIZONTALLY ON VISE AS SHOWN FOR CHECKERING. TURN PART EACH TIME PORTION IS CHECKERED.
8
BREAK ALL SHARP EDGES AFTER CHECKERING WITH EMERY CLOTH & BUFF. PART NOW READY FOR FITTING.

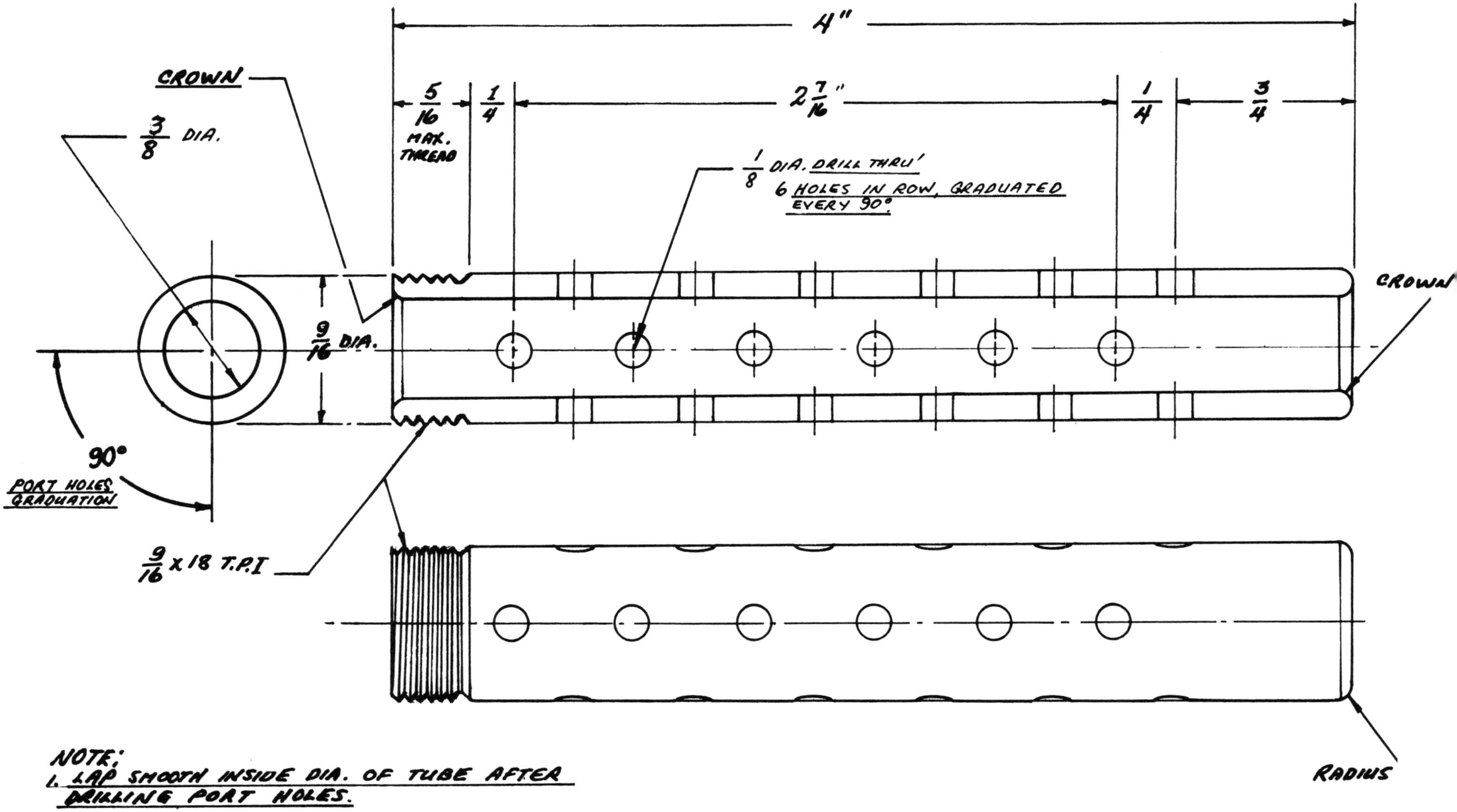

BARREL EXTENSION TUBING

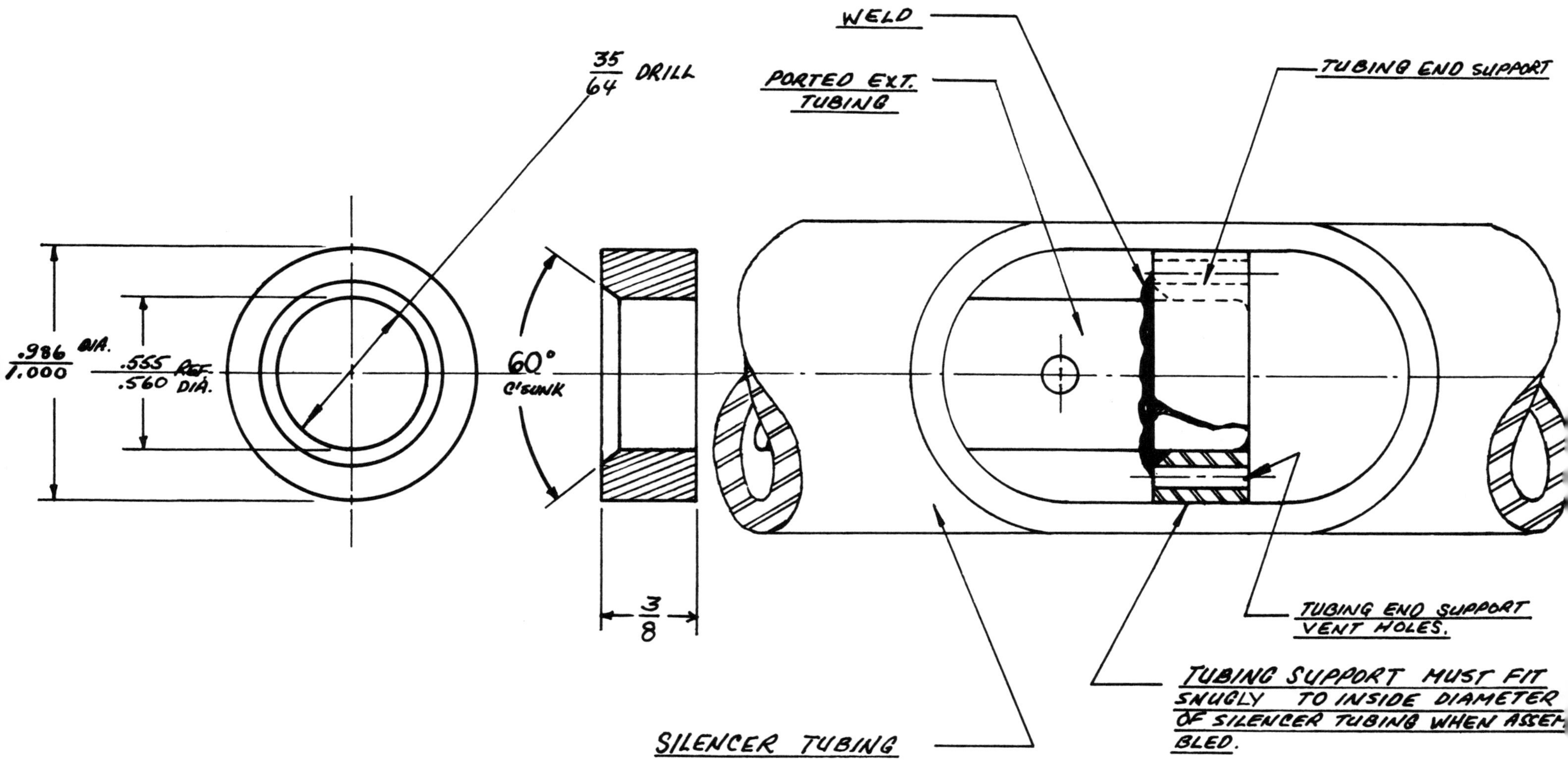

EXTENSION TUBING END SUPPORT

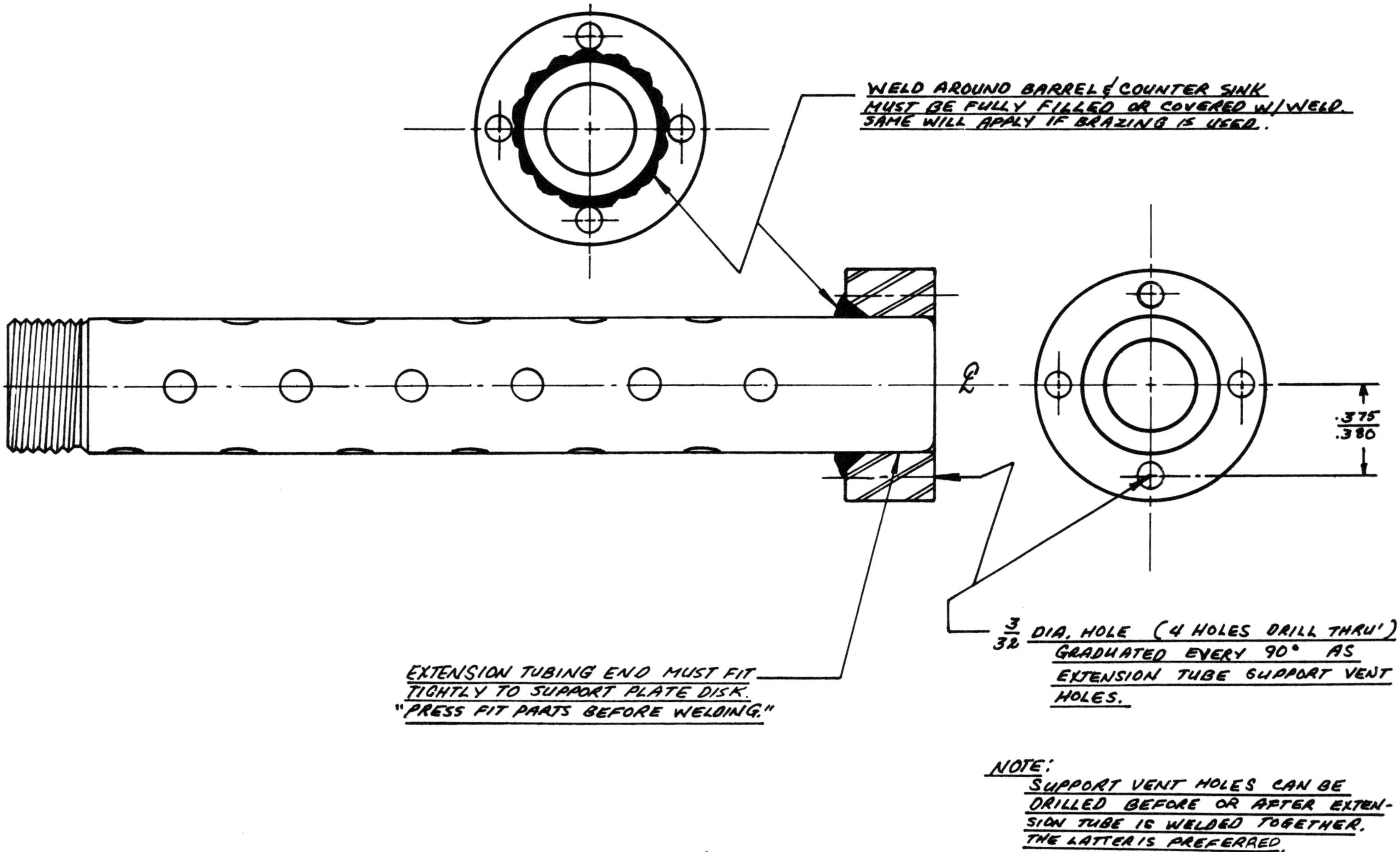

EXTENSION TUBING & SUPPORT ASSEMBLY

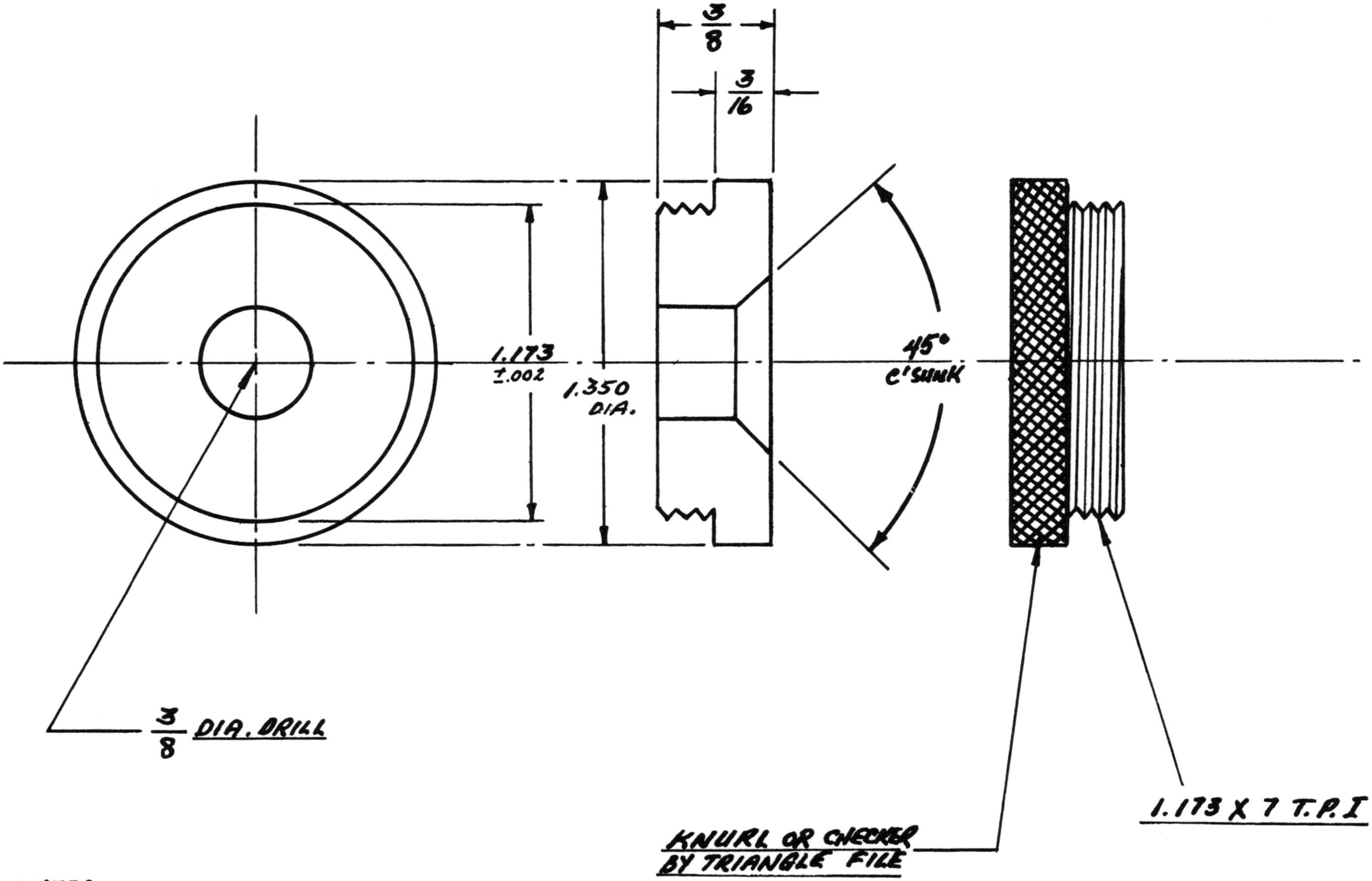

NOTES:

1. BREAK EXTERNAL EDGES AFTER THREADING & KNURLING.
2. POLISH SMOOTH INSIDE DIA. (EXIT HOLE) WITH FINE EMERY CLOTH & LAP.

SILENCER TUBING END CAP

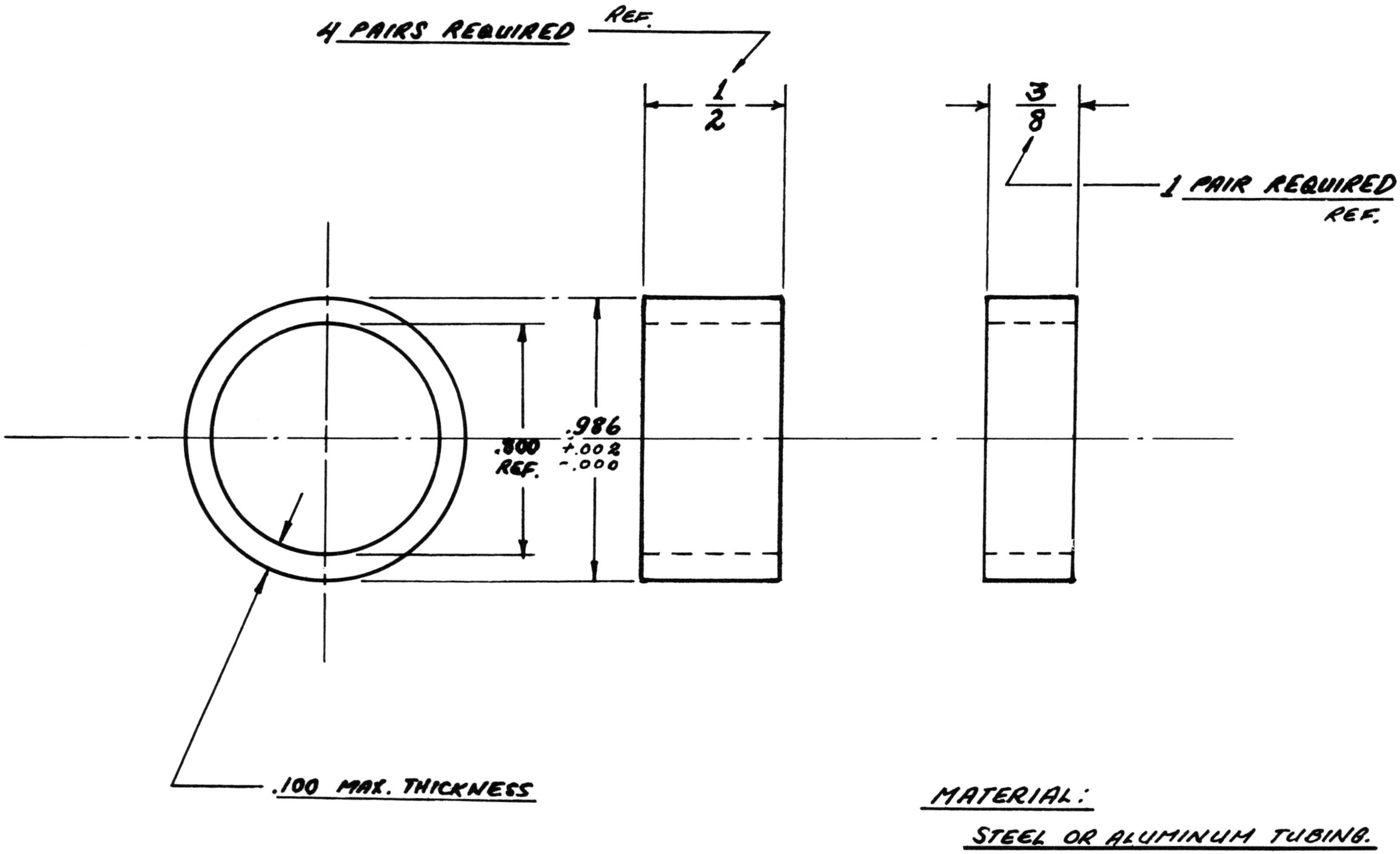

BAFFLE SPACER

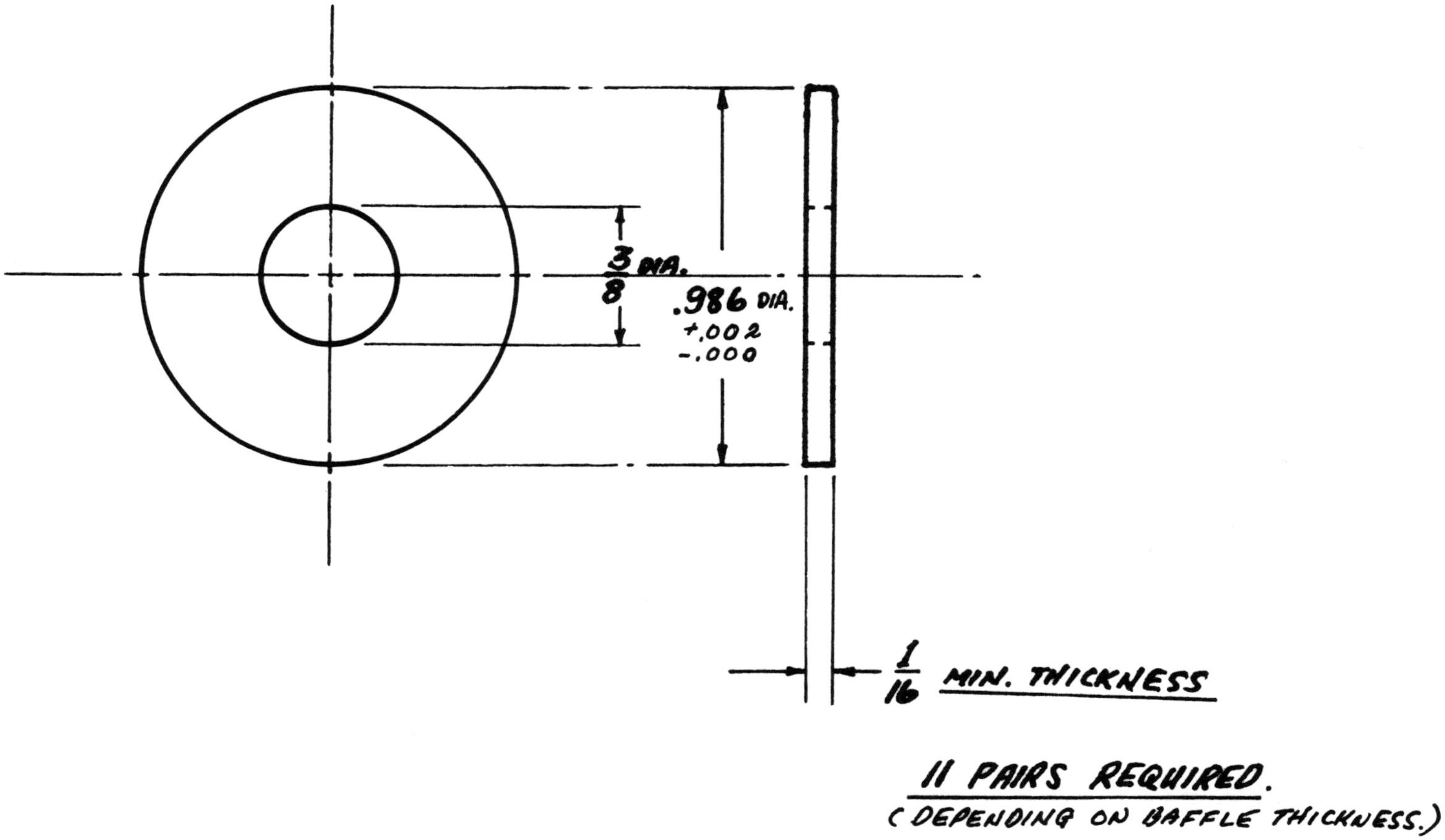

BAFFLE SUPPORT

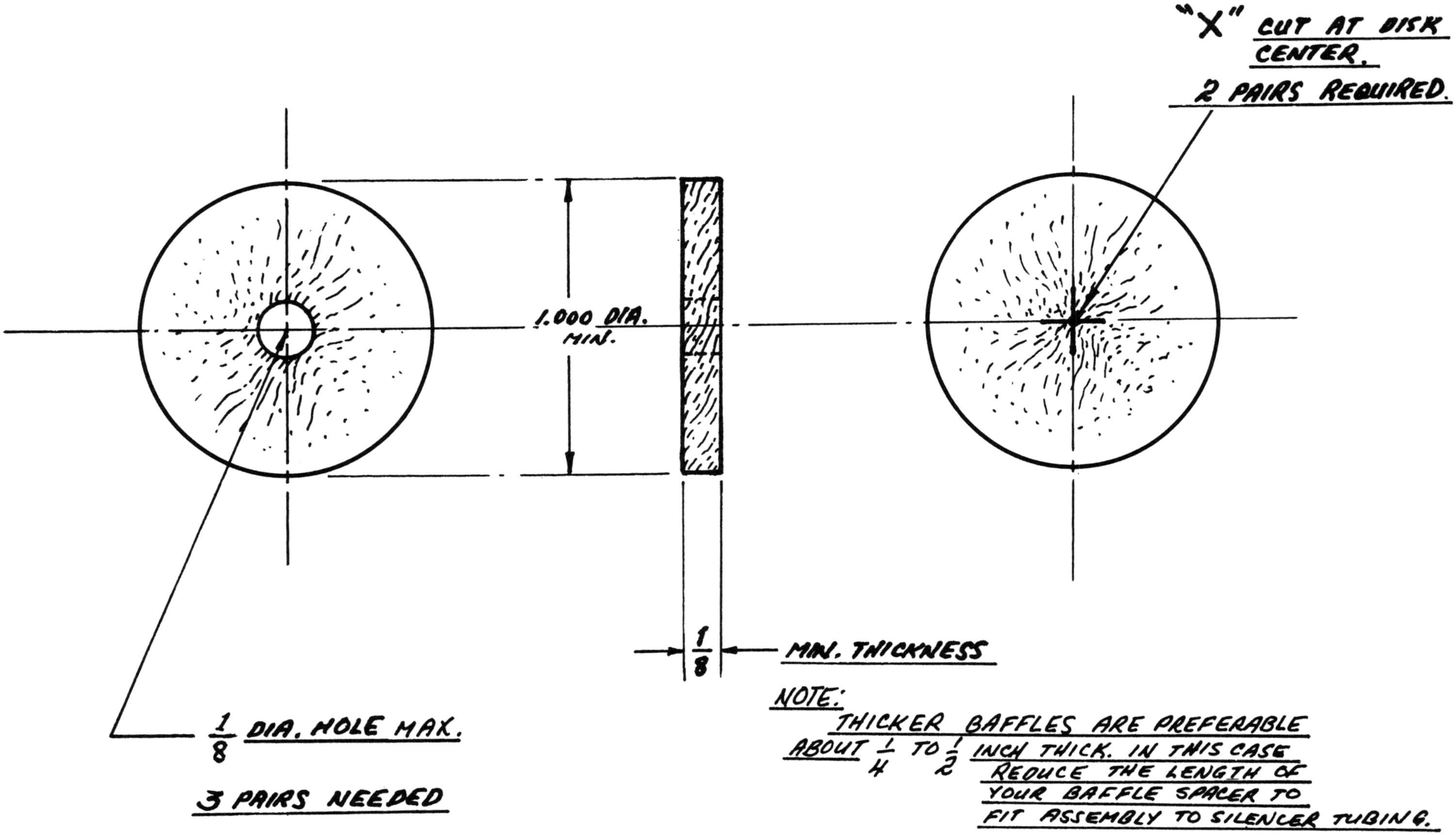

RUBBER BAFFLE DISK

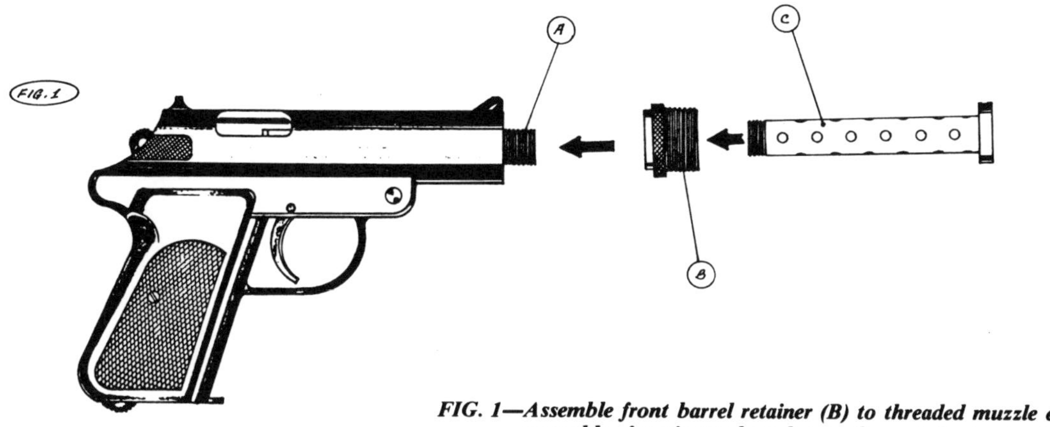

FIG. 1—Assemble front barrel retainer (B) to threaded muzzle end of barrel (A). After assembly of retainer to barrel screw the ported extension tubing (C) to the threaded hole of the retainer tightly.

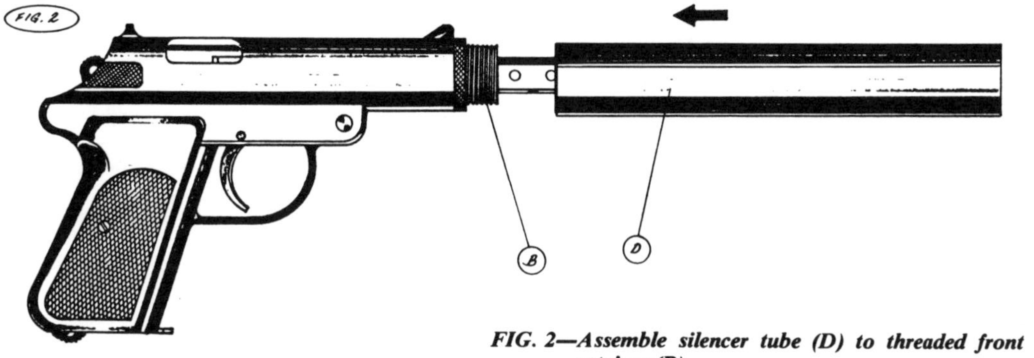

FIG. 2—Assemble silencer tube (D) to threaded front extension of the barrel retainer (B).

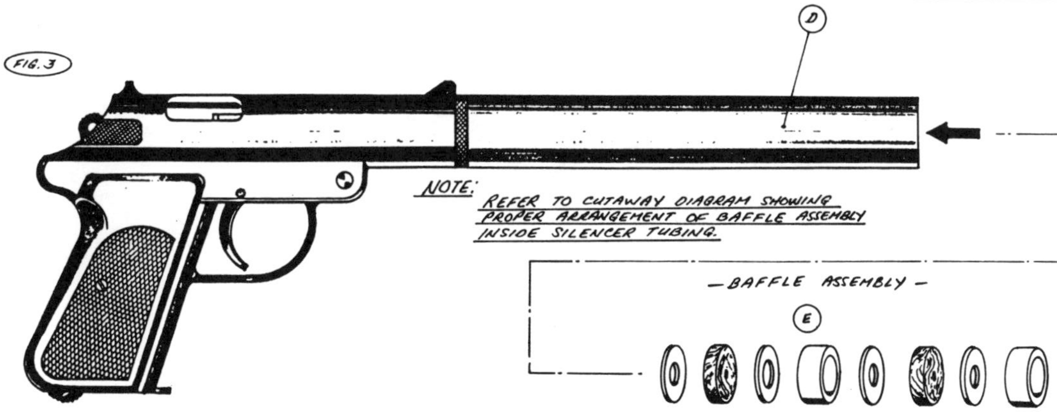

FIG. 3—Insert and assemble all baffle units (E) inside silencer tube (D). Refer to cutaway diagram showing the proper arrangement of the baffle assembly.

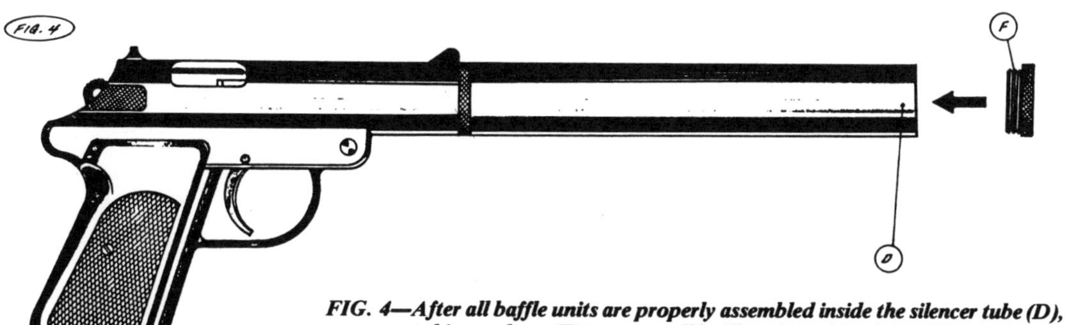

FIG. 4—After all baffle units are properly assembled inside the silencer tube (D), screw on the silencer tubing end cap (F) to secure all baffle units inside tubing. Silencer is fully assembled and ready for use. Reverse procedure for disassembly.

VII. BAFFLE CUTTER TOOL

Unlike other silencer designs currently on the market, my silencers need no fiberglass or similar packing in the expansion chamber. Their advanced design, particularly the ported barrel support, make such packing unnecessary. This means you will not have to disassemble the silencer to repack the asbestos every few dozen rounds.

You will, however, have to change the rubber baffle disks occasionally. The life span of a good hard rubber baffle is normally about 200 rounds. I have designed the cutter tool shown in the following drawings to facilitate the quick and uniform cutting of the rubber baffles.

The cutter tubing can be made from the same tubing that forms the outer tube of the silencer. This is ideal. Its length can be from 6 to 8 inches. The cutting end is chamfered 5 degrees inside and 10 degrees outside. If you don't have a grinder or lathe to do the chamfering with, you'll have to use a half-round file and alot of patience to form this cutting edge.

Once both surfaces are chamfered as required, sharpen the cutter's edge with rough emery cloth, then switch to a fine emery cloth for a final polish.

If the tubing is steel of known analysis, it may be heat treated for long life.

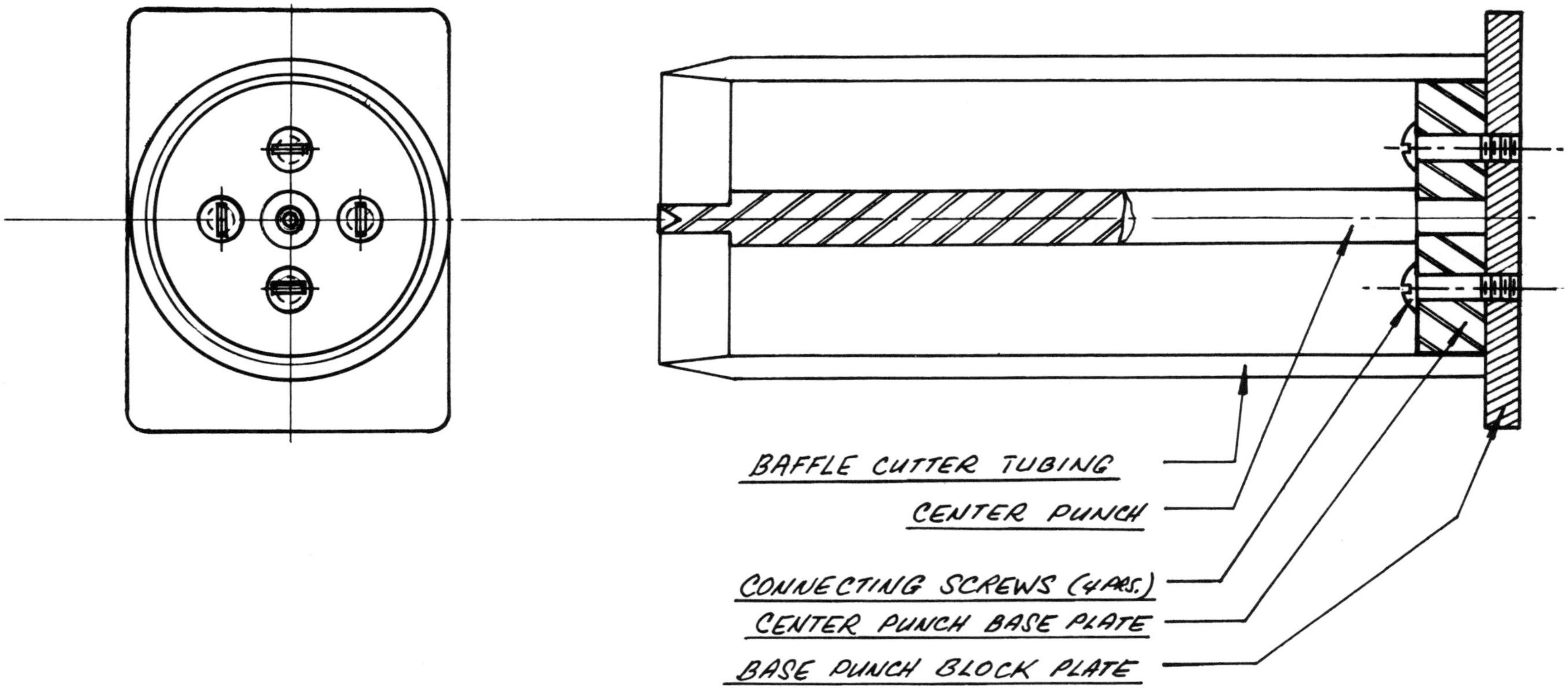

- IMPROVISED BAFFLE CUTTER TOOL -

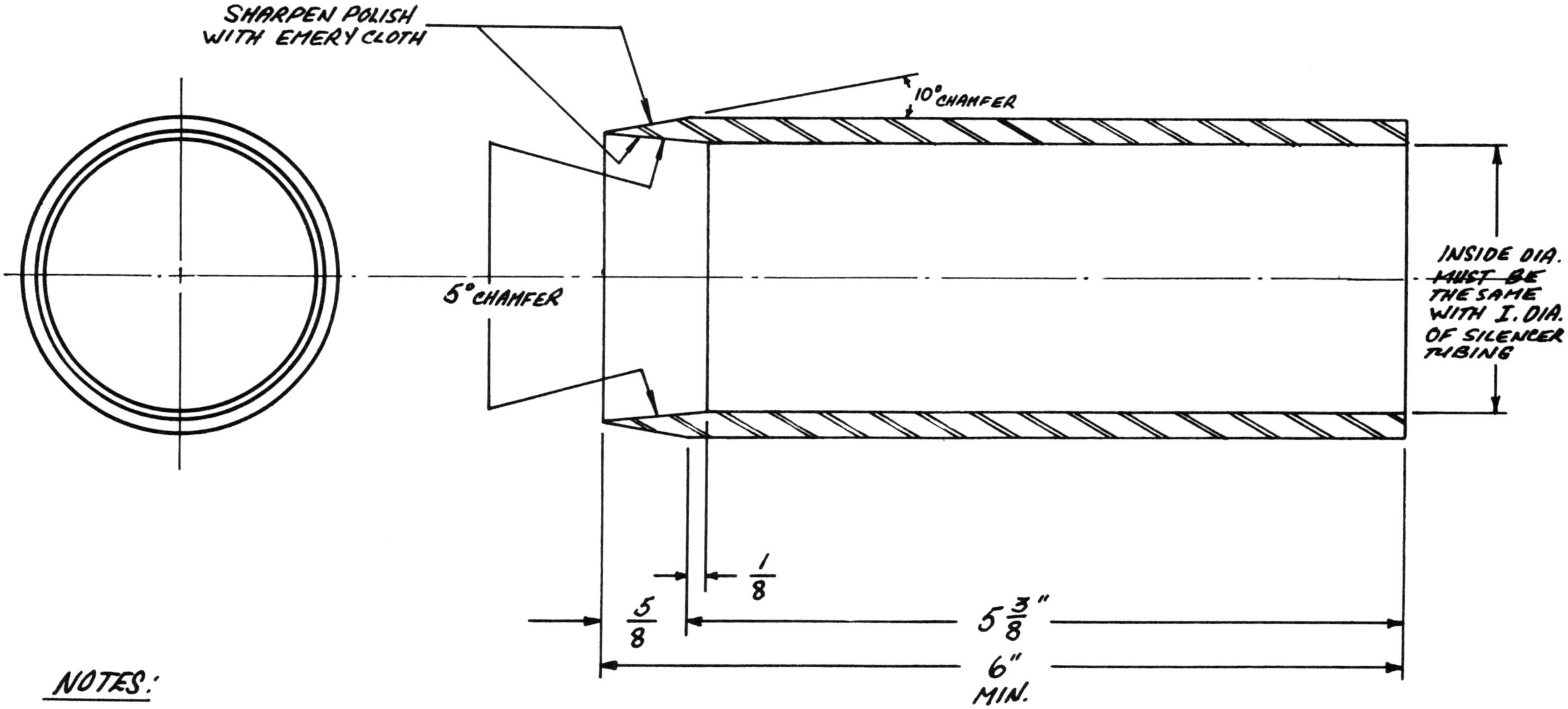

NOTES:

1. INSIDE CHAMFER CAN BE DONE BY USING ROUND GRINDER OR FILING IT BY HALF ROUND FILE.
2. OUTSIDE CHAMFER CAN BE DONE BY TURNING PART TO LATHE OR FILING IT WITH MILL BASTARD FILE.
3. AFTER CHAMFERING OPERATION, SHARPEN POINT WITH SEMI-FINE EMERY CLOTH.

BAFFLE CUTTER TUBING

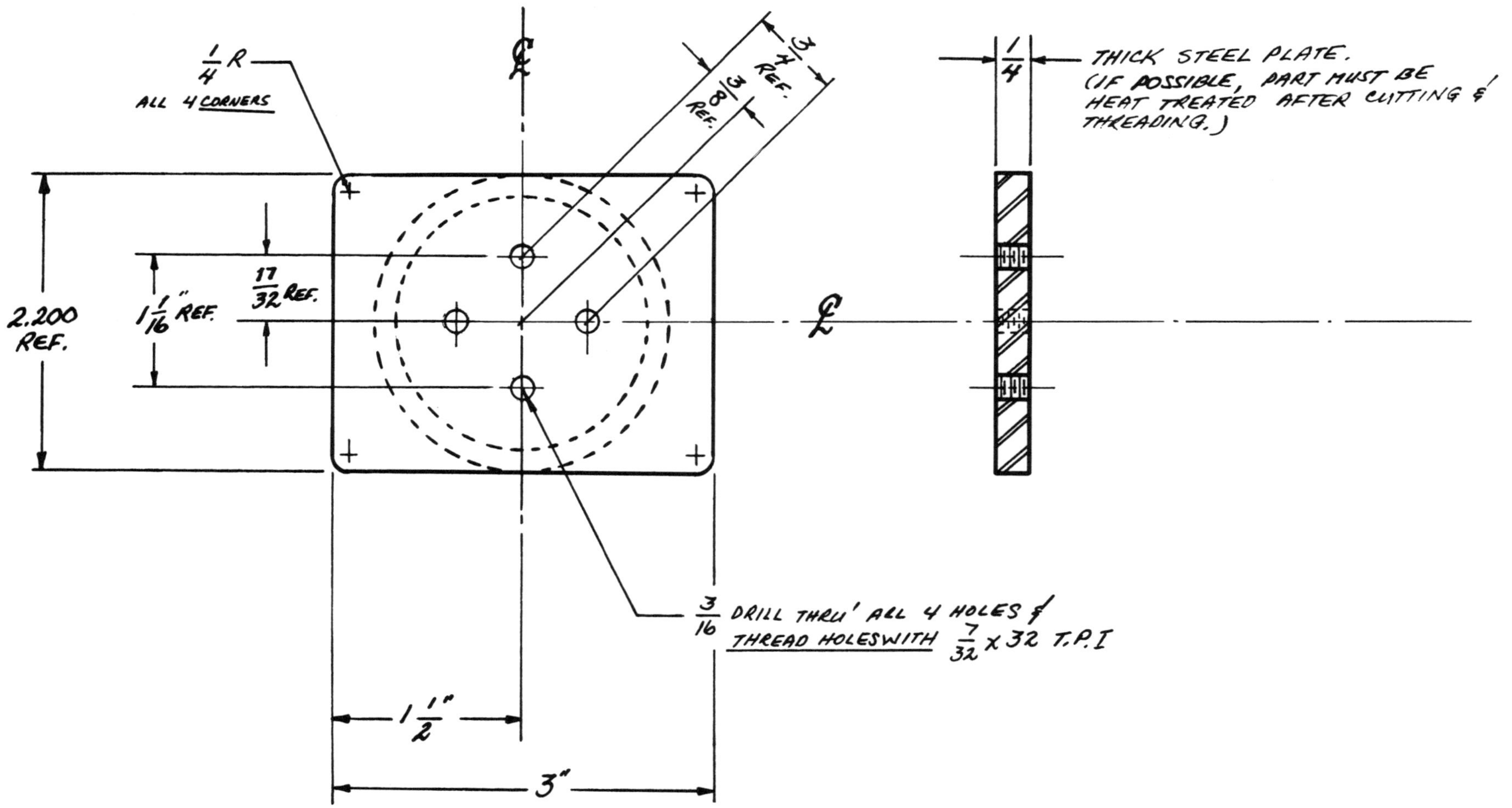

BASE PUNCH BLOCK PLATE

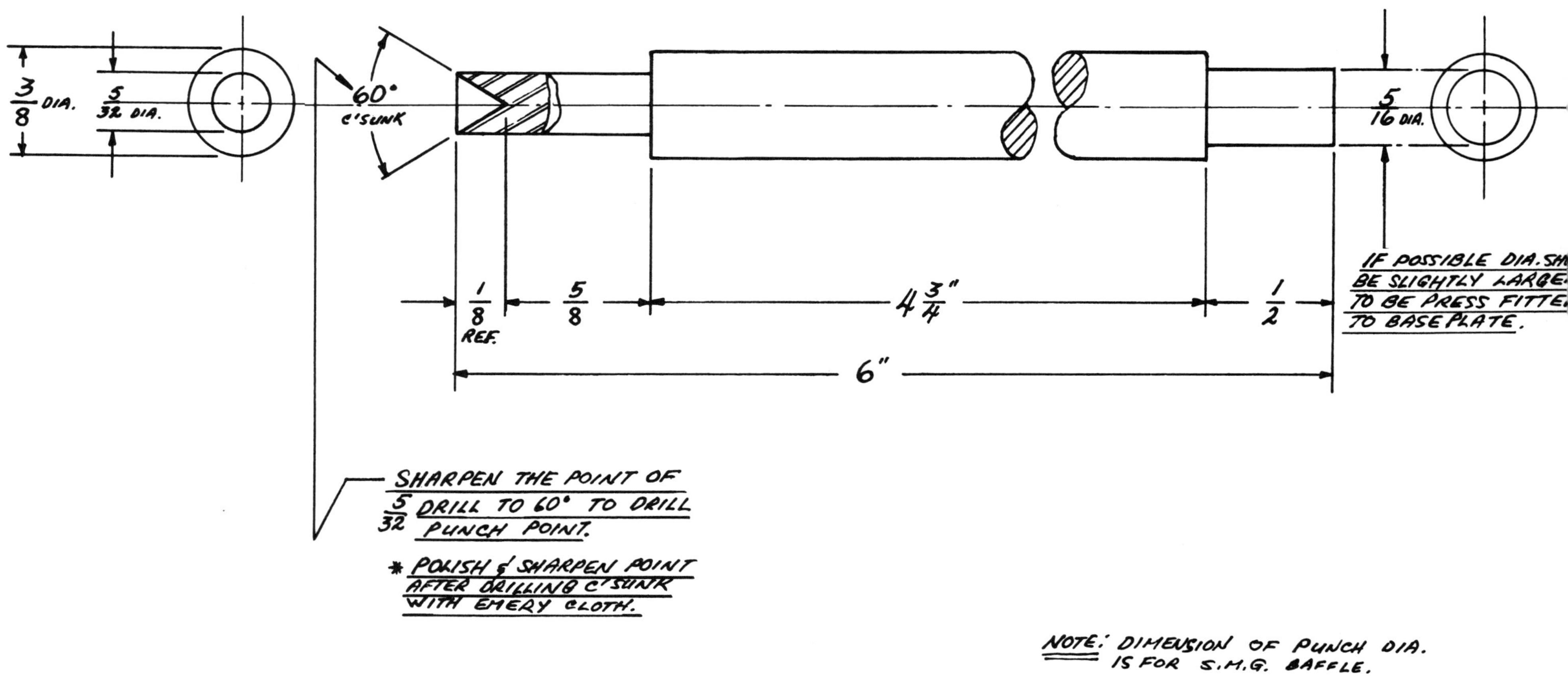

CENTER PUNCH

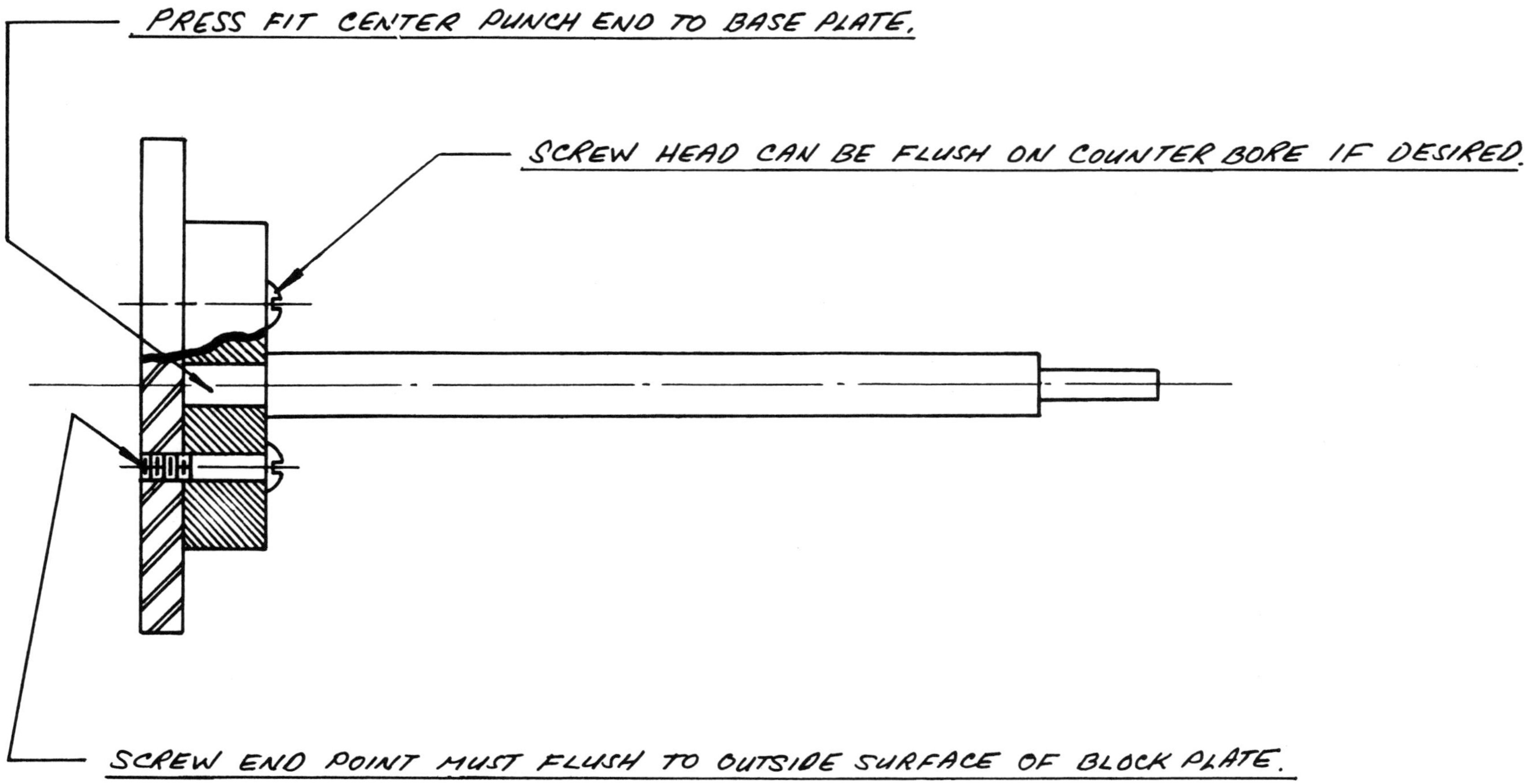

-CENTER PUNCH ASSEMBLY-

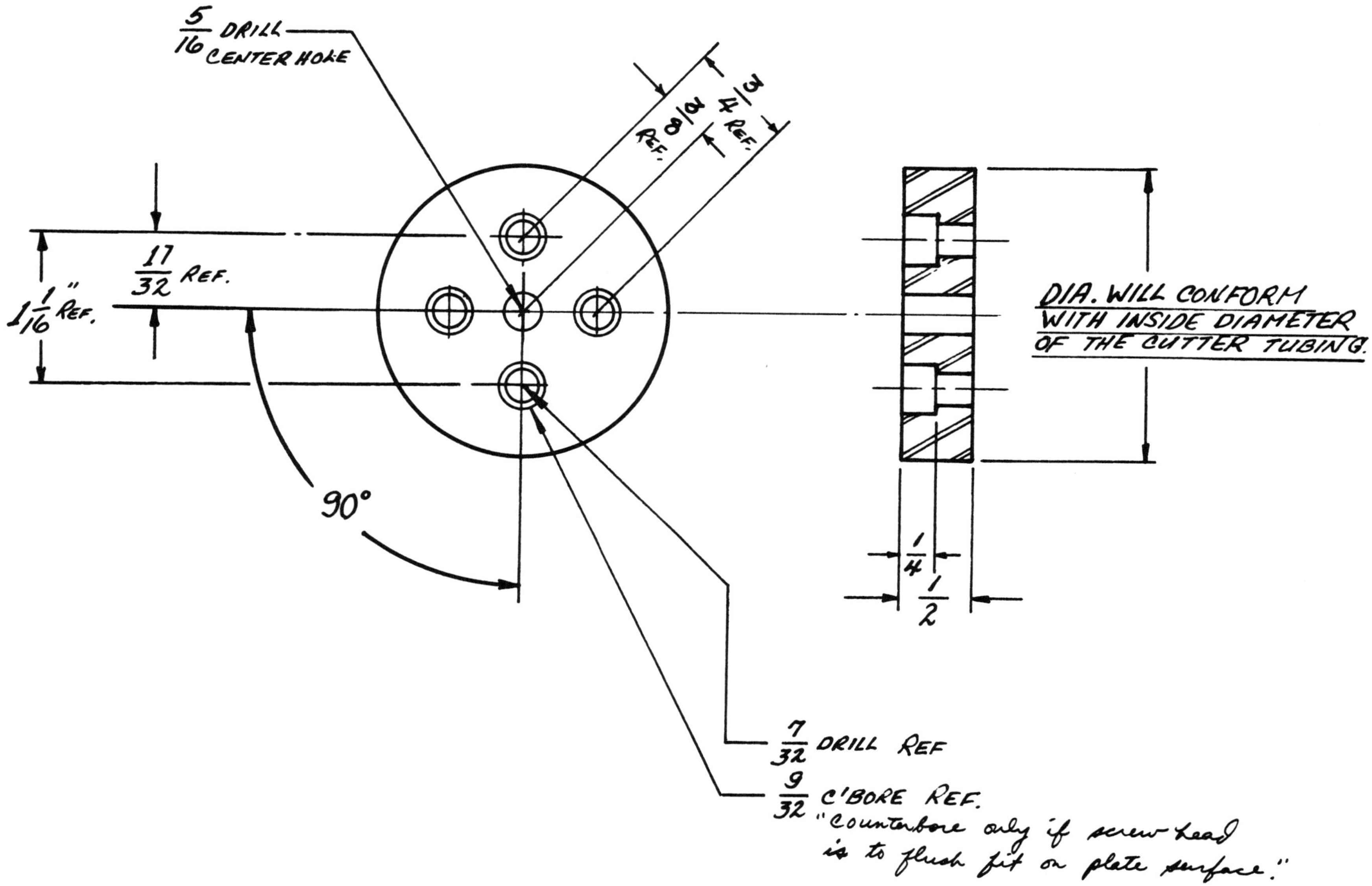

CENTER PUNCH BASE PLATE